Simon Aurand

A Contribution to Friction in Timber Connections

BAND 40
Karlsruher Berichte zum Ingenieurholzbau

Herausgeber
Karlsruher Institut für Technologie (KIT)
Holzbau und Baukonstruktion
Prof. Dr.-Ing. Philipp Dietsch

A Contribution to Friction in Timber Connections

by
Simon Aurand

KIT Scientific Publishing

Karlsruher Institut für Technologie
Holzbau und Baukonstruktion

A Contribution to Friction in Timber Connections

Zur Erlangung des akademischen Grades eines Doktors der Ingenieurwissenschaften (Dr.-Ing.) von der KIT-Fakultät für Bauingenieur-, Geo- und Umweltwissenschaften des Karlsruher Instituts für Technologie (KIT) genehmigte Dissertation

von Simon Aurand, M.Sc.

Tag der mündlichen Prüfung: 6. Juni 2024
Referent: Prof. Dr.-Ing. Hans Joachim Blaß
Korreferent: Prof. Dr.-Ing. Thomas K. Bader

Impressum

KIT Scientific Publishing

Karlsruher Institut für Technologie (KIT)
KIT Scientific Publishing
Straße am Forum 2
D-76131 Karlsruhe

KIT Scientific Publishing is a registered trademark
of Karlsruhe Institute of Technology.
Reprint using the book cover is not allowed.

www.bibliothek.kit.edu/ksp.php | E-Mail: info@ksp.kit.edu | Shop: www.ksp.kit.edu

Print on Demand 2025 – Gedruckt auf FSC-zertifiziertem Papier

ISSN 2511-6312
ISBN 978-3-7315-1399-5
DOI 10.5445/KSP/1000176783

Für Jonathan

Abstract

Timber connections with inclined screws are state-of-the-art and widely used. They can be used for timber-to-timber and steel-to-timber connections or in combination with system connectors. Due to their inclination regarding the shear plane, inclined screws are practically only loaded in tension, as the share of axial load in the total load-carrying capacity far exceeds the share of lateral load. This load distribution is due to different stiffness values in the axial and lateral directions of the fasteners (the difference is about one order of magnitude). Therefore, connections with inclined screws fully utilise the screws' capacity if designed correctly.

Due to equilibrium reasons, there is a normal force in the shear plane, pressing the connected parts together. This force activates friction in the shear plane, and an additional portion of load can be transferred. Therefore, the load-carrying capacity of connections with inclined screws can be increased because of friction. This work investigates if the friction between connecting parts can be increased with surface modifications. Subsequently, tests are conducted with connections with inclined screws and increased friction in the shear plane.

Friction can be classified into different types, including static and sliding friction. The coefficient of friction μ represents the dimensionless relationship between frictional and normal force (with frictional force being parallel and normal force being perpendicular to the surface). Contrary to popular belief, a specific material has no fixed friction coefficient. Friction is a system effect influenced by multiple factors, such as material properties, surface texture, sliding speed, and environmental conditions.

Different modification processes are investigated for their feasibility, efficiency, and potential optimisation in manufacturing. Tests with densified veneer wood (DVW) indicate transferability to materials like steel or aluminium. Simpler manufacturing processes yield less consistent surface quality, while more complex surface modifications lead to consistently high surface quality. Milled surfaces are very effective, especially those with longitudinal and transverse grooves. Embossed surfaces and certain coatings show promise, while coatings with larger mineral grains are not recommended.

Friction tests with the manufactured surfaces and softwood show increased friction coefficients compared to untreated surfaces of DVW or aluminium. Notably, prominent, protruding features are not necessary for high coefficients; an embossed surface with an inverse pyramid pattern performs very well. The results emphasise the importance of even surface structures for an even load distribution across the shear plane.

Tests with connections with modified steel or DVW surfaces show increased load-carrying capacities. Connections with inclined screws (mainly axially loaded) and bolts inserted perpendicular to the shear plane (mainly laterally loaded) are performed. Particularly the tests with the inclined screws show a significant increase in load-carrying capacity. The stiffness increases more with inclined screws than with bolts inserted perpendicular to the shear plane. The results highlight that surface modifications increase the load-carrying capacity more than the stiffness and emphasise the advantages of connections with inclined screws.

An analytical model predicts the load-carrying capacity well for connections with inclined screws and different surfaces and friction coefficients, screw lengths, and numbers. Additional numerical models simulate the deformation behaviour and provide information about the stresses in the connectors.

Tests in changing environmental conditions and wood moisture content indicate a slight decrease in the load-carrying capacity over time. Stiffness increases slightly for all tests in both service classes. Evaluated creep factors are small in a conditioned environment (service class 1) but increase in an unconditioned environment (service class 2).

Overall, the results provide insights into the short-term and long-term behaviour of connections with inclined screws and increased friction in the shear plane. An appropriate surface modification increases the friction coefficient in the shear plane, thus increasing the load-carrying capacity.

Kurzfassung

Verbindungen im Ingenieurholzbau mit geneigten Schrauben entsprechen dem aktuellen Stand der Technik und sind weit verbreitet. Sie können für Holz-Holz- und Stahl-Holz-Verbindungen oder in Verwendung mit Systemverbindern eingesetzt werden. Aufgrund ihrer Neigung zur Scherfuge werden geneigte Schrauben praktisch nur auf Zug beansprucht, da der Anteil der Axialbeanspruchung an der Gesamttragfähigkeit den Anteil der Abscherbeanspruchung weit übersteigt. Dies ist auf die unterschiedlichen Steifigkeitswerte bei Axial- und Abscherbeanspruchung der Schrauben zurückzuführen (der Unterschied beträgt etwa eine Größenordnung). Daher wird bei Verbindungen mit geneigten Schrauben die Tragfähigkeit der Schrauben bei richtiger Auslegung optimal ausgenutzt.

Aus Gleichgewichtsgründen wirkt in der Scherfuge eine Normalkraft, die die verbundenen Teile zusammenpresst. Diese Normalkraft aktiviert Reibung in der Scherfuge, so dass ein zusätzlicher Anteil an Last übertragen werden kann. Daher kann die Tragfähigkeit von Verbindungen mit geneigten Schrauben durch Reibung erhöht werden. In dieser Arbeit soll untersucht werden, ob die Reibung zwischen zwei Bauteilen durch entsprechende Oberflächenmodifikationen erhöht werden kann. Anschließend werden Versuche mit Verbindungen mit geneigten Schrauben und erhöhter Reibung in der Scherfuge durchgeführt.

Die Reibung kann in verschiedene Arten eingeteilt werden, darunter Haft- und Gleitreibung. Der Reibungskoeffizient μ ist das dimensionslose Verhältnis zwischen Reib- und Normalkraft (wobei die Reibkraft parallel und die Normalkraft senkrecht zur Oberfläche wirkt). Entgegen der weitverbreiteten Meinung besitzt ein bestimmtes Material keinen festen Reibungskoeffizienten. Reibung ist ein Systemeffekt, der von mehreren Faktoren beeinflusst wird, z.B. von den Materialeigenschaften, der Oberflächenbeschaffenheit, der Gleitgeschwindigkeit und den Umgebungsbedingungen.

In dieser Arbeit werden verschiedene Modifizierungsverfahren auf ihre Machbarkeit, Effizienz und mögliche Optimierung in der Fertigung untersucht. Versuche mit Kunstharzpressholz (KPH) zeigen eine Übertragbarkeit auf Materialien wie Stahl oder Aluminium. Einfachere Modifizierungsverfahren führen zu einer weniger gleichmäßigen

Oberflächenqualität, während komplexere Oberflächenmodifikationen zu einer gleichbleibend hohen Oberflächenqualität führen. Gefräste Oberflächen sind sehr effektiv, insbesondere solche mit Längs- und Quernuten. Geprägte Oberflächen und bestimmte Beschichtungen sind vielversprechend, während Beschichtungen mit größeren Mineralkörnern nicht zu empfehlen sind.

Reibversuche mit den hergestellten Oberflächen und Nadelholz zeigen erhöhte Reibungskoeffizienten im Vergleich zu unbehandelten Oberflächen aus KPH oder Aluminium. Dabei ist hervorzuheben, dass markante, hervorstehende Merkmale nicht notwendig sind für hohe Reibbeiwerte; eine geprägte Oberfläche mit einem Negativabdruck eines Pyramidenmusters erzielt sehr gute Reibbeiwerte. Die Ergebnisse unterstreichen die Bedeutung einer gleichmäßigen Oberflächenstrukturen für eine gleichmäßige Lastverteilung in der Scherfuge.

Versuche mit Verbindungen mit modifizierten Stahl- oder KPH-Oberflächen zeigen erhöhte Tragfähigkeiten. Es werden Verbindungen mit geneigten Schrauben (primär auf Zug beansprucht) und mit senkrecht zur Scherfuge angeordneten Passbolzen (primär auf Abscheren beansprucht) durchgeführt. Insbesondere die Versuche mit den geneigten Schrauben zeigen eine signifikante Erhöhung der Tragfähigkeit. Die Steifigkeit nimmt bei geneigten Schrauben stärker zu als bei den Versuchen mit den Passbolzen. Die Ergebnisse zeigen, dass Oberflächenmodifikationen die Tragfähigkeit stärker erhöhen als die Steifigkeit und unterstreichen in diesem Zusammenhang die Vorteile von Verbindungen mit geneigten Schrauben.

Ein analytisches Rechenmodell bestimmt die Tragfähigkeit für verschiedene Oberflächen und Reibungskoeffizienten, Schraubenlängen und -anzahlen gut. Zusätzliche numerische Modelle simulieren das Verformungsverhalten und geben Auskunft über die Spannungen in den Verbindern.

Tests unter wechselnden klimatischen Bedingungen und Holzfeuchten zeigen eine leichte Abnahme der Tragfähigkeit im Laufe der Zeit. Die Steifigkeit nimmt bei allen Versuchen in beiden Nutzungsklassen leicht zu. Die ausgewerteten Kriechfaktoren sind in einer kontrollierten Umgebung (Nutzungsklasse 1) gering, nehmen aber in einer unkontrollierten Umgebung (Nutzungsklasse 2) zu.

Insgesamt geben die Ergebnisse Aufschluss über das Kurzzeit- und Langzeitverhalten von Verbindungen mit geneigten Schrauben und erhöhter Reibung in der Scherfuge. Eine geeignete Oberflächenmodifikation erhöht den Reibungskoeffizienten in der Scherfuge und steigert damit die Tragfähigkeit.

Acknowledgement

This thesis was written while working as a research assistant at KIT Timber Structures and Building Construction. Excerpts of the results were published in different journals [1, 2, 3] and presented at various conferences [4, 5, 6].

I want to express my special thanks to my supervisor, Professor Hans Joachim Blaß, who provided me with the opportunity to do my doctorate at his department and was always available to discuss the topic and offer guidance. I would also like to thank Professor Thomas K. Bader, my second reviewer, for his interest in my work and constructive dialogue.

I am grateful to all my colleagues, past and present, for their valuable input and discussions. I would also like to thank the staff at the Karl-Möhler-Holzbaulaboratorium for their assistance in manufacturing the test specimens and conducting the experiments. Additionally, I would like to thank all the students who contributed significantly to this work, either through their Bachelor's and Master's theses or as student assistants.

Finally, I would like to express my heartfelt gratitude to my family, who have always supported me, especially during turbulent times.

Karlsruhe, November 2024
Simon Aurand

Contents

Abstract . . . . **i**

Kurzfassung . . . . **iii**

Acknowledgement . . . . **v**

1 Introduction **1**
- 1.1 Motivation 1
  - 1.1.1 Research problem 1
  - 1.1.2 Problem description 2
  - 1.1.3 Connections with dowel-type fasteners 3
  - 1.1.4 Research questions 7
  - 1.1.5 Research programme 8
- 1.2 Structure of the thesis 9

2 Friction **11**
- 2.1 Introduction 11
- 2.2 Literature review 15
- 2.3 Evaluation of the results from literature 25
- 2.4 Comparison to values given in standards 35
- 2.5 Parameter study based on own experiments 37
  - 2.5.1 Test setup and execution 37
  - 2.5.2 Results 41
- 2.6 Conclusion 44

3 Surface modification **47**
- 3.1 Introduction 47
- 3.2 Notching 48
- 3.3 Embossing 50
- 3.4 Punching 51
- 3.5 Profile milling 53
  - 3.5.1 Pyramid pattern 53
  - 3.5.2 Circular pattern 54

3.5.3 Horizontal grooves 55
3.5.4 Scale pattern 55
3.6 Belt grinding 56
3.7 Sandblasting 56
3.8 Brushing 57
3.9 Coating 57
3.9.1 Two-component adhesive 57
3.9.2 Epoxy adhesive tape 58
3.9.3 Grip tape 60
3.10 Conclusion 60

4 Friction tests 63
4.1 Introduction 63
4.2 Test setup and execution 63
4.3 Results and discussion 68
4.3.1 Untreated 68
4.3.2 Notched 71
4.3.3 Embossed 73
4.3.4 Punched 75
4.3.5 Milled 78
4.3.6 Sanded 84
4.3.7 Sandblasted 84
4.3.8 Brushed 84
4.3.9 Coated 85
4.4 Statistical evaluation 89
4.5 Characteristic values 92
4.6 Conclusion 96

5 Connection tests 99
5.1 Introduction 99
5.2 Connections with laterally loaded fasteners 99
5.2.1 Test programme and setup 99
5.2.2 Results and discussion 101
5.2.3 Summary 106
5.3 Connections with inclined fasteners 107
5.3.1 Test programme 107
5.3.2 Test setup 108
5.3.3 Results and discussion 110
5.3.4 Main beam to secondary beam tests 126
5.4 Summary and conclusions 132

6 Modelling and comparison to experimental results **135**
6.1 Introduction and objectives 135
6.2 Analytical model 136
6.2.1 Design equations 136
6.2.2 Comparison with experimental results 141
6.2.3 Monte Carlo simulation 144
6.2.4 Comparison with characteristic values 149
6.2.5 Influence of the insertion angle 150
6.3 2D FE-Model 153
6.3.1 Structure of model 153
6.3.2 Characteristics of springs 154
6.3.3 Validation of springs 160
6.3.4 Comparison with experimental results 162
6.3.5 Summary 2D 170
6.4 3D FE-Model 170
6.4.1 Structure of model 170
6.4.2 Modelling of axially loaded screws 171
6.4.3 Calibration of *cohesive surface* 172
6.4.4 Comparison with experimental results 173
6.4.5 Summary 3D 177
6.5 Conclusion 178

7 Long-term behaviour **181**
7.1 Introduction 181
7.2 Connections with inclined fasteners 184
7.2.1 Duration-of-load tests 184
7.2.2 Residual load-carrying capacity 195
7.2.3 Influence of shrinking and swelling 201
7.2.4 Summary 203
7.3 Connections with laterally loaded fasteners 204
7.3.1 Duration-of-load tests 204
7.3.2 Residual load-carrying capacity 211
7.3.3 Summary 215
7.4 Comparison with results from the literature 216
7.5 Conclusion 219

8 Conclusions and Recommendations **221**
8.1 Conclusions 221
8.2 Recommendations 225

Bibliography **227**

Acronyms and symbols **239**

A Appendix **243**
A.1 Friction tests 243
A.2 Connection tests 259
A.3 Long-term tests 277
A.4 Modelling 281

1 Introduction

1.1 Motivation

1.1.1 Research problem

The concept of friction seems so inconspicuous, but it enormously impacts everyday life. In short, nothing works without friction. Pedestrian movement is impossible without friction between the soles of shoes and the ground. As much as friction is relied upon unconditionally in everyday life, it is barely relied upon in the design of timber structures. Contact between components such as a wall and a floor is generally regarded as frictionless. For timber connections, however, research has shown that even minor adjustments can transfer additional forces via friction [7]. This research has since been adopted in the codes for designing timber connections. However, only the friction coefficient between wood and wood may be considered. This consideration does not assess whether the contact is only between wood and wood or if the contact is actually between wood and another material or surface, which in combination may lead to a higher coefficient of friction.

In 1940, Gaber [8] conducted the first tests to increase the friction in the shear plane of timber-to-timber connections. By adding sand between the wood, the friction coefficient could be increased. Almost 30 years later, Möhler & Maier [9] also performed tests intending to increase the friction in timber-to-timber connections by adding sand in the shear plane. Both concluded that the friction coefficient can be increased when just the right amount of sand is added but that the risk of failure is just too high to further pursue this method. As both only performed tests to determine the coefficient of friction and not tests with timber-to-timber connections, no conclusion was made about the effect of the increased friction coefficient on timber-to-timber connections. Current research, however, did show a positive influence of friction in the shear plane on the load-carrying capacity of connections [10].

Therefore, the following points were analysed herein in more detail:

- Generally, the determination of friction coefficients between two different materials and surfaces, and methods for increasing the friction between these two surfaces
- Specifically, the consideration of friction and its influence on connections when calculating the load-deformation behaviour

1.1.2 Problem description

Coefficient of friction and increasing friction

The first significant publications on the subject of friction in timber structures date back to the middle of the 20th century (e.g. [8, 11, 12]). All these publications have in common the material wood, used to determine coefficients of friction in various combinations. Tests were also often carried out with wood and steel. However, what distinguishes all these publications from one another is the selection or, rather, the specification of the parameters analysed. Of a total of 3500 values found in the literature, a maximum of approx. 850 can be compared with each other at any one time. The choice of individual parameters often also significantly impacts other parameters. For example, the test setup indirectly determines how high the contact pressure can be. This leads to sometimes contradictory results in the literature. It is clear that, despite decades of research, new aspects still need to be considered. For this reason, correlations between various parameters influencing the coefficient of friction between wood and wood-based materials are being investigated. It also appears that methods for increasing the friction between two components and/or connectors still need to be explored. For this purpose, herein different methods of surface modifications are to be analysed, manufactured, tested, and compared with each other.

Design of connections considering the coefficient of friction

The great potential of accounting for friction in the design of connections outside timber structures has been recognised. In slip-resistant connections in steel structures, the load is transferred via friction in the shear plane of the connected steel plates. Preloaded bolts assure the contact between the steel plates (but do not transfer load via embedment). For timber structures in earthquake-prone regions, so-called *slip-friction connectors* can be used. Like the slip-resistant connections, steel plates are connected with bolts. These connectors dissipate energy by allowing significant deformations and friction between the moving parts [13, 14]. For temporary timber structures in the form of scaffolding, friction can be applied to guarantee positional stability [110].

Nevertheless, especially for permanent timber structures, there is great potential in taking friction into account in the analytical design to increase the load-carrying capacity of connections, thus making them more effective. Calculation models are required to account for the influence of friction. Analytical and numerical models should be able to estimate the load-carrying capacity and stiffness. Tests must also be carried out to verify the accuracy of these models. Various combinations of surfaces and fasteners are analysed in the tests. When talking about connections, two types of shear connections are herein differentiated:

- Connections with fasteners, inserted perpendicular to the shear plane, that are mainly loaded laterally
- Connections with fasteners, inserted inclined at an angle to the shear plane, that are mainly loaded axially

1.1.3 Connections with dowel-type fasteners

1.1.3.1 Laterally loaded fasteners

Standard timber connections are designed with dowel-type fasteners inserted perpendicular to the shear plane (at a right angle $\varepsilon = 90°$). Under load parallel to the shear plane, the fasteners are primarily loaded perpendicular to their longitudinal axis. This work refers to these connections with *laterally loaded fasteners*. The load transfer in connections with such laterally loaded fasteners primarily occurs via the direct (compressive) contact between the fastener and the wood. The governing parameters are the wood's embedment strength and the fastener's bending strength. The embedment strength of the wood depends on the grain direction and is denoted by f_h. The bending strength of the fastener is defined by its yield moment M_y. The analytical design of connections with laterally loaded fasteners is based on the theory of Johansen [15]. Backed by test results, Johansen determined analytical equations for dowels and bolts. Different types of failure occur depending on the selection of the geometric parameters, i.e. the thickness of the wood and the length and diameter of the fastener. The results of Johansen's equations are usually given per fastener and shear plane. To optimally utilise the connection, the aim is to achieve failure of the fastener with two plastic hinges per shear plane (failure mode F). See exemplarily Equation 1.1 of a single-shear timber-to-timber connection with two plastic hinges:

$$R = \sqrt{\frac{2 \cdot \beta}{1+\beta}} \cdot \sqrt{2 \cdot M_{\mathrm{y}} \cdot f_{\mathrm{h},1} \cdot d} \tag{1.1}$$

with R lateral resistance per shear plane
β ratio of embedment strength $f_{\mathrm{h},2}$ / $f_{\mathrm{h},1}$
M_{y} yield moment
$f_{\mathrm{h},1}$ embedment strength of member 1
d diameter of the fastener

In addition, certain boundary conditions must be adhered to when designing connections with fasteners inserted perpendicular to the shear plane that are loaded parallel to the shear plane. Minimum spacing and end and edge distances ensure that premature failure due to the splitting of the wood does not occur. This is the only way to ensure plastic hinges form in the fastener (besides reinforcement of the timber member). To account for non-uniform load distribution in connections with multiple fasteners, a reduced effective number of fasteners in load and grain direction has to be considered (see n_{ef} in Eurocode 5 [114]).

Significance of friction

For all failure modes where the fastener's axis is inclined to the shear plane (ε < 90°), normal forces in the fastener occur. If the fastener can transmit normal forces, a force F_{n} perpendicular to the shear plane occurs for equilibrium reasons. The force F_{n} activates friction, leading to an increased load-carrying capacity [7]. The Johansen model for connections with bolts already included a share due to friction. The coefficient of friction μ determined by Johansen varied between 0.4–1.4 (therefore, it is assumed that the static coefficient of friction was used). Möller [16] transferred the theory to nailed connections but did not consider an increasing effect on the total load due to friction (or disregarded it, as smooth nails only have a low withdrawal capacity). Meyer [17], in turn, took friction into account and named this load-increasing effect the *rope effect*. In this analogy, the fastener is regarded as a rope pulling the two connected timber parts together, which explains the name itself.

In conclusion, however, Meyer recommended neglecting the rope effect of the nail due to the large scatter of frictional properties. Here, it is essential to note that the rope effect regards the friction in the shear plane and not the friction between the wood and the fastener. Two decades later, Ehlbeck [18] confirmed this decision but spoke in favour of further research in this field, as he saw the great potential of the rope effect, especially in connections with threaded nails with high withdrawal resistance. In this work, *rope effect* always refers to the effect of additional forces parallel to the shear plane due to friction between the two adjoining members.

Not until 2004, when the new edition of the German standard DIN 1052 [106] for the design of timber structures was released, an increase in load-carrying capacity due to friction was considered in the design of timber connections. When using a coefficient of friction of μ = 0.25, the following Equation 1.2 results.

$$R = \sqrt{\frac{2 \cdot \beta}{1+\beta}} \cdot \sqrt{2 \cdot M_{\mathrm{y}} \cdot f_{\mathrm{h},1} \cdot d} + 0.25 \cdot F_{\mathrm{ax,R}} \tag{1.2}$$

with	R	lateral resistance per shear plane
	β	ratio of embedment strength $f_{h,2}$ / $f_{h,1}$
	M_y	yield moment
	$f_{h,1}$	embedment strength of member 1
	d	diameter of the fastener
	$F_{ax,R}$	axial resistance of the fastener

However, the portion of the rope effect in the load-carrying capacity is limited to a percentage of the capacity, according to Johansen's theory. The percentage increases with increasing withdrawal capacity of the fastener. For dowels this is 0%, as the smooth shank is not considered to transfer normal forces to the timber. For screws, the percentage is 100%, as the withdrawal capacity of screws is the highest among dowel-type fasteners.

Current research has shown that even for dowels with a smooth shank, non-negligible proportions of the total load-carrying capacity occur due to friction [19, 20, 21]. Furthermore, in [10] the use of a frictional component in the analytical design of laterally loaded connections is confirmed and emphasised to account for the further development of the withdrawal capacity of dowel-type fasteners. Nevertheless, significant displacements are needed to fully account for the additional frictional share (confirmed in [22]).

1.1.3.2 Inclined fasteners

Modern timber connections are designed with fasteners (i.e. fully threaded screws) inserted inclined at an angle $\varepsilon < 90°$ to the shear plane. Under load parallel to the shear plane, the fasteners are primarily loaded axially, in a direction longitudinal to their axis. Bejtka & Blaß [7] showed experimentally and analytically that even at minimal angles of inclination, the share of axial load in the total load-carrying capacity far exceeds the share of lateral load. This is due to different stiffness values in the axial and lateral direction of the fasteners (the difference is about one order of magnitude). At an angle $\varepsilon = 45°$, practically the entire load is transferred by axial loading of the screw, and the share from lateral loading can be neglected. This work refers to these connections as *inclined fasteners* or *axially loaded fasteners*. The load transfer in connections with inclined screws primarily takes place via the interlocking of the screw's thread and the wood. The design, therefore, consists of three failure types. That is 1) failure of the bond between the screw's thread and the wood when reaching the withdrawal capacity, leading to the screw being pulled out of the wood. 2) Failure of the screw itself when reaching the tensile load-carrying capacity, leading to the screw rupture. 3) Failure of the wood adjacent to the screw's head when reaching the head pull-through capacity, leading to the screw being pulled through the wood on the side of the screw head.

Significance of friction

Bejtka & Blaß [7] showed that additional forces could be transmitted via friction between the timber parts even with a slight inclination of the screws. The Johansen model was extended to account for inclined screws under lateral load. With $\varepsilon = 90°$, the failure modes for laterally loaded screws are decisive, and the extended Johansen model results in Equation 1.2 (for failure mode F of a single-shear timber-to-timber connection).

For an angle $\varepsilon < 90°$ and neglecting the lateral share due to embedment, a force equilibrium is achieved as shown in Figure 1.1. Normal forces in the fastener lead to forces perpendicular to the shear plane that, in turn, activate friction. An additional load component parallel to the shear plane due to friction can be considered for the total load-carrying capacity.

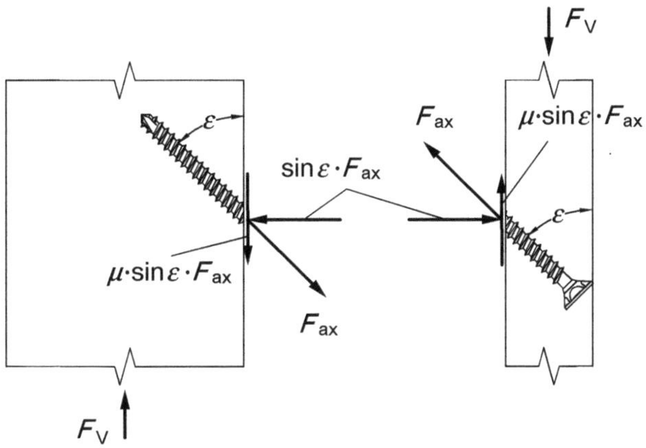

Figure 1.1: Force diagram of acting forces in a connection with inclined screws.

The total load-carrying capacity F_V of the connection is thus calculated as the sum of the parallel part of the force in the inclined fastener and the friction force in the shear plane, see Equation 1.3:

$$F_V = F_{ax} \cdot (\cos\varepsilon + \mu \cdot \sin\varepsilon) \tag{1.3}$$

with	F_V	load-carrying capacity of the connection
	F_{ax}	axial capacity of the fastener
	ε	insertion angle
	μ	friction coefficient

1.1.4 Research questions

Key questions

What types of surface modification for materials such as steel, aluminium, and similar are available that can be used to increase the coefficient of friction in the shear plane? Can this high coefficient of friction be reflected in the calculation and subsequent installation situation of timber connections?

Small scale friction tests

What suitable surface modifications for steel, aluminium, and similar are available? Can a consistent surface finish always be guaranteed (reproducibility)? Is the surface treatment industrially feasible? Moreover, particularly important: how does the surface behave in friction tests with wood, what coefficient of friction can be determined, and how can it be determined?

Full scale connections tests

How do connections with significantly higher friction in the shear plane behave? Can increases in the coefficient of friction be considered linearly in the design of connections? Is friction still present in connections even after a long service life, and to what extent?

1.1.5 Research programme

Against the background of the research questions, a very detailed literature research on friction concerning wood and timber structures was conducted at the beginning of the work. A database of test results with around 3500 entries was created. The test results from the literature were extensively analysed and evaluated regarding the various influencing parameters. The findings from the literature research were confirmed, and some unanswered questions were clarified experimentally with a self-conducted parameter study on the coefficient of friction. At the same time, various surface treatment options were considered and analysed for their suitability for connections in timber structures. As a result of these studies, friction tests were carried out with various modified surfaces (in steel, aluminium, and densified veneer wood) and softwood.

An analytical model was developed to determine the load-carrying capacity of connections using the determined friction coefficients. This calculation model was continuously validated and adapted with results from tests of timber-to-timber and steel-to-timber connections with laterally loaded fasteners (inserted perpendicular to the shear plane) and inclined screws (mainly loaded axially). Different surface modifications and different types and numbers of fasteners were considered. In further steps, numerical models were investigated to identify the stiffness and deformation behaviour of the connection (two-dimensional model) and the stresses in the connector itself (three-dimensional model) in addition to the load-carrying capacity.

1.2 Structure of the thesis

Figure 1.2 shows the structure of the thesis. The thesis follows a clear path from reviewing and understanding friction to subsequent surface modifications. Influencing parameters on friction are determined and discussed. The surface modifications are mainly viewed against the background of industrial production. In friction tests, corresponding friction coefficients are determined, and surface modifications for later applications are defined. Connections manufactured with the chosen surface modifications are subsequently tested, and their load-carrying capacity is determined. An analytical and numerical calculation model is being developed and constantly compared with the test results. Parallel to this, long-term tests are carried out with connections with modified surfaces. Finally, the most important findings of the work are summarised, and recommendations for application and further research are given.

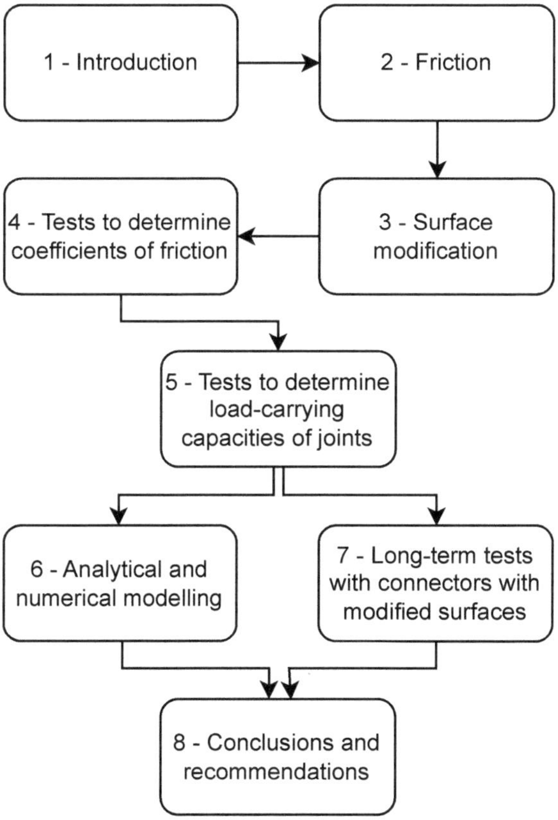

Figure 1.2: Structure of the thesis.

2 Friction

2.1 Introduction

Friction has great practical significance in daily life, such as wheels on the road, hands on the steering wheel, friction in brakes or clutches, stating some of the most prominent examples. Others include friction between the feet and the shower floor, between the shoes and the sidewalk or between paper and rubber rollers in any office printer, which transport the paper due to friction [23]. In general, countless mechanisms and processes are based on friction. This underlines why friction is indispensable.

Leonardo da Vinci (1452–1519) was one of the first scientists to perform systematic tests to study the phenomenon of friction. For the first time, he introduced the coefficient of friction as the ratio between frictional force and normal force. Almost 200 years later, the French physicist Guillaume Amontons (1663–1705) rediscovered the findings of da Vinci regarding friction and formulated two laws stating that [24]:

1. the frictional force is independent of the surface area
2. the frictional forces are directly proportional to the applied load

Another 100 years later, Charles Augustin de Coulomb (1736–1806), another French physicist, extended Amontons' laws by adding that [24]:

3. the frictional force has a maximum value
4. static friction is higher than kinetic friction

As Coulomb already discovered, there are different types of friction: sliding friction, rolling friction, friction between solids contacted by fluids and static friction or blocking [23]. Friction can be partly explained by a fundamental principle of physics, which states that for every force, there is an equal opposed force (actio = reactio). So, the frictional force is often the reaction force dictated by physics [23].

If two bodies lying on top of each other are moved relative to each other, friction occurs in the contact area between these two bodies; see Figure 2.1. One of the two bodies remains resting until a specific force $F_{f,stat}$ is exceeded. From that point on, the body moves, pushed by the force $F_{f,kin}$. The magnitude of the force $F_{f,stat}$ and $F_{f,kin}$ depend on the coefficient of friction (COF or friction coefficient). The friction coefficient μ describes the dimensionless relation between the friction force F_f and the normal force F_n which keeps the two bodies in contact. So the static coefficient of friction μ_s corresponds to the maximum force $F_{f,stat}$ which must be exceeded to trigger macroscopic movement between the two bodies, while the kinetic coefficient of friction μ_k corresponds to the force $F_{f,kin}$ which is required to maintain movement between the two bodies.

The coefficient of friction μ is the dimensionless ratio of the frictional force F_f parallel to the contact surface, to the force F_n perpendicular to the contact surface.

- Static friction: the static frictional force $F_{f,stat}$ must be overcome to move a resting body. This critical force is proportional to the normal force F_n:

$$F_{\mathrm{f,stat}} = \mu_{\mathrm{s}} \cdot F_{\mathrm{n}} \quad (2.1)$$

- Sliding friction: the kinetic frictional force $F_{f,kin}$ is the force that keeps the body moving after the critical force is overcome:

$$F_{\mathrm{f,kin}} = \mu_{\mathrm{k}} \cdot F_{\mathrm{n}} \quad (2.2)$$

with μ_s static coefficient of friction
μ_k kinetic coefficient of friction

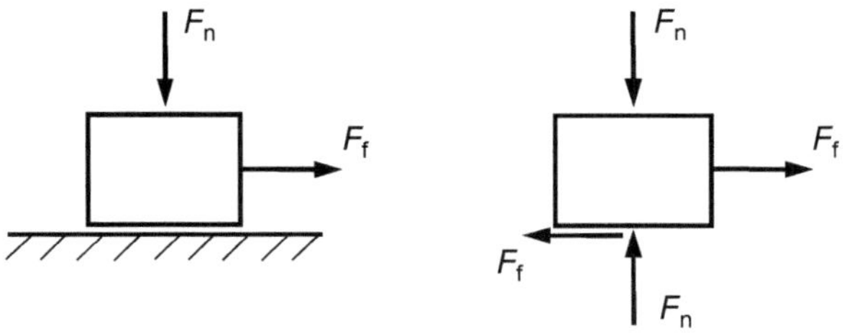

Figure 2.1: A body on the plane, which is stressed by normal and frictional forces.

Contrary to general belief, a specific material has no specific friction coefficient. Friction is always a system effect influenced by many factors. The coefficient of friction is always individual for a system consisting of two materials, their mechanical properties and their surface texture, the sliding speed and type of movement, the geometry of the contact, as well as surrounding conditions [23]. This is affirmed by the fact that friction is an energy-dissipating process. A resting material dissipates no energy and can therefore not have a friction coefficient [23]. Therefore, if friction coefficients are presented, the materials and the testing conditions should always be included.

As can be seen in Figure 2.2, there are usually three different ways a system for determining the coefficient of friction behaves during testing [23]:

- Diagram 1 shows a sliding system where the force required to start the motion F_a is greater than the force required to maintain the motion F_b. Therefore, the force F_a is used to calculate the static coefficient of friction μ_s and the force F_b is used to calculate the kinetic coefficient of friction μ_k.
- Diagram 2 shows the force-time-curve when stick-slip occurs. Stick-slip can usually be eliminated by changing the stiffness of the tested system, the load, or the sliding speed. When stick-slip occurs in a system undergoing a friction test, it is common not to report a coefficient of friction. Instead, "stick-slip behaviour" should be reported as result.
- Diagram 3 shows the force-time-curve for a pairing without pronounced adhesion peak. The force F_c would be used to calculate the static coefficient μ_s and the force F_b would be used to calculate the kinetic coefficient μ_k. Sometimes the sliding friction force increases or decreases with time. This indicates an unstable system. In this case it is suggested to specify the coefficient of friction at prescribed time/sliding intervals.

Stick-slip is generally a dynamic, cyclic process in which two contacting surfaces oscillate between a stick and a slip phase. In the stick phase, the two surfaces do not move and are held in place by the static frictional force. There is finite relative motion in the slip phase, and the kinetic frictional force maintains this motion. When force is applied to move the surface of one component relative to the other, there is resistance to movement until sufficient force is applied to overcome the static friction. When this happens, one component breaks away and rapidly accelerates until it reaches the same speed as the other moving component. At this point, there is no relative motion between the two surfaces, and static friction occurs. Assuming the original force is maintained, the cycle repeats.

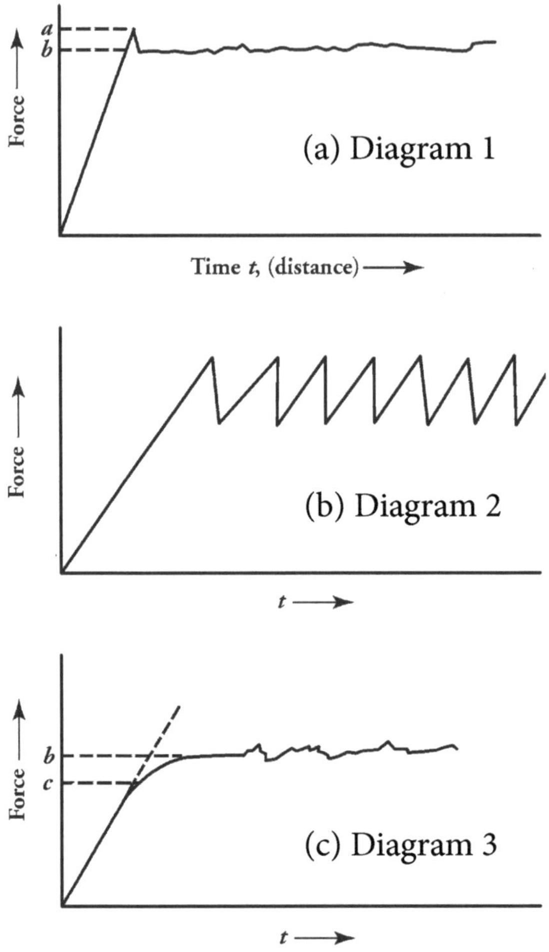

Figure 2.2: Typical behaviour of different systems in friction tests [23].

Static and kinetic friction forces are complicated and can be time and relative velocity-dependent. As the speed of the component decreases, the amplitude of the stick-slip process is greater because there is more time for the static friction to maximise, so more force is required to cause movement, resulting in a greater displacement or amplitude. Conversely, as the speed of the component increases, the amplitude decreases because there is less time for static friction to build up. At some point, as the speed increases, the amplitude is so small that the stick-slip process ceases. In addition, surface roughness can affect the amplitude and speed of the cycle and even whether there is a pronounced stick-slip behaviour. Stick-slip can usually be eliminated by changing the stiffness, load or speed of the tribosystem[1] [23, 25].

2.2 Literature review

In the following, the findings in the literature regarding the coefficient of friction in tests with wood on wood and wood on steel are presented. These results were partly presented in [6]. The reason for the experiments of the presented literature can be roughly divided into four motives and are given in chronological order of the appearance of the research:

- the influence of friction on wear and tear of machine parts (1950s)
- basic research on friction and the friction coefficient (1960s – present)
- the influence of friction on wood processing (1970s)
- the influence of friction on the structural design of connections and structures (2000 – present)

Wear and tear of machine parts

The first relevant tests on friction started with the problem of wear and tear of machine parts. The need to understand the processes and mechanisms behind wear and tear led researchers to perform fundamental experiments. As friction was (and still is) primarily responsible for the wear of machine parts and, therefore, of great importance, the need for experiments arose. Fundamental experiments were performed by Bowden [26] and Atack & Tabor [11] in the late 1950s. Friction tests with wood on steel were mainly performed to determine the kinetic friction coefficient, as the wear and tear happened with moving machine parts. They discovered that the frictional force is composed of at least two parts. One part is due to the adhesion of the paired surfaces, and one part is due to the deformation of the surfaces, which they called the "ploughing term".

[1] A tribosystem is a system consisting of at least two contacting bodies and any environmental factors affecting their interaction.

$$F_s = F_a + F_d \tag{2.3}$$

with F_s sliding frictional force

F_a tangential force required to overcome interfacial adhesion

F_d tangential force involved in deforming ("ploughing term")

The effect of the "ploughing term" is visualized in [27] and can be seen in Figure 2.3, where a sharp point is progressively pushed into a surface and dragged along. The sharpness of the point affects the degree of penetration into the surface and, therefore, the magnitude of the "ploughing" contribution to the frictional force.

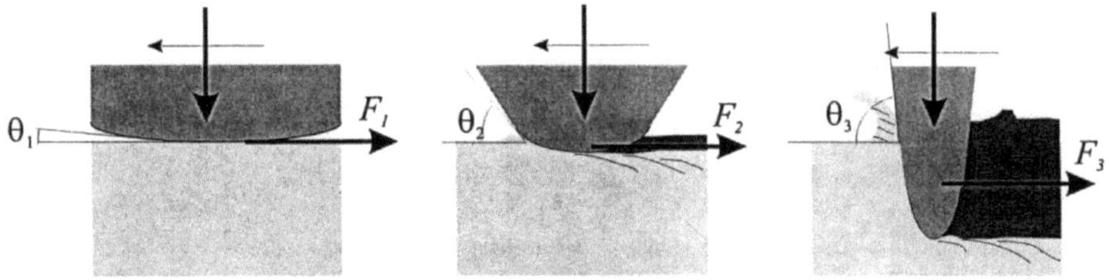

Figure 2.3: Effect of the slider geometry on the "ploughing part" of the tangential force [27].

Basic research

With a change of the materials to wood on wood, the first systematic experiments were performed by Stošić [12] in 1959. In 9000 tests, he systematically investigated the four parameters: wood species, surface quality, fibre direction and contact pressure. He also differentiated between two test setups. His conclusions include that the friction coefficient decreases with the surface's fineness and increases with the surface's roughness, respectively. Also, the friction coefficient is greater for tests where the fibre direction in the shear plane was parallel than for tests with end grain. He recorded the greatest friction coefficients with the lowest contact pressure. Finally, he concluded that the test setup with an inclined plane is unsuitable for determining the coefficient of friction.

In 1961, McLaren & Tabor [28] determined the kinetic coefficient of friction for dry and wet lignum vitae (*Guaiacum*) on steel. Their interest was in using lignum vitae as underwater bearings because of its self-lubricating properties. In conclusion, friction coefficients were independent of contact pressure, and the adhesion term in Equation 2.3 could be verified.

The influence of the contact pressure and the moisture content of the wood was investigated by Murase [29] in tests with wood on wood and wood on steel. He could confirm Amontons-Coulomb's law that the coefficient of friction is (widely) independent of the contact pressure. However, he concluded that the friction coefficient is significantly dependant on the moisture content.

The most recent research on the basics of friction between wood and steel was done by Kuwamura [30]. He primarily investigated the influence of high contact pressure (> 1 MPa). His findings concluded that increasing contact pressure decreases the friction coefficient in the cases of "incomplete friction" or rough steel surfaces. The term "incomplete friction" was introduced by Kuwamura and "refers to cases where part of the movement of the pultruded plate is consumed by shear-yield deformation of the wood" [30].

Wood processing

With the increased use of wood as building material and the manufacture of engineered wood products, friction's influence on wood processing became more relevant. Understanding and controlling friction in the manufacturing process could reduce cost and time and improve production quality. Experiments were therefore carried out in the context of wood processing, and tests were mainly carried out with wood and steel. McKenzie & Karpovich [31] performed friction tests with wood on steel, as this significantly influenced the cutting and planing of timber. They determined static and kinetic friction coefficients. Their newest finding was that the fibre direction of the wood had only a low impact on the friction coefficient. They also confirmed former researchers by stating that the moisture content and increasing contact pressure influence the friction coefficient.

Lemoine et al. [32] investigated in detail the sliding friction of wood on steel in the context of wood machining. They specifically investigated the influence of the following wood characteristics: 1) moisture content, 2) density, 3) extractive content, 4) annual ring position, and 5) early wood/late wood. Overall, their results confirm the previous experiments. Thus, hardly any influence of the fibre direction (parallel or perpendicular) could be determined. In tests with end grain, however, the coefficient of friction was higher. They were also unable to determine any influence of the density on the coefficient of friction. Based on this, McMillin et al. [33] investigated the influence of different lubricants. The lubricants had different effects depending on whether the wood was dry or wet. The lubricant increased friction in oven-dry samples, whereas in saturated samples, the lubricant decreased friction. Also of interest in wood machining is the influence of temperature on the coefficient of friction. This was also investigated by McMillin et al. [34], and he concluded that the coefficient of friction decreases with increasing temperature.

Other studies on the influence of (mainly sliding) friction on the machining of wood include the influence on the wear of the tool used [35], the influence on the feed rate during cutting [36], the influence of different materials (different steel types and sapwood/heartwood) [37] and lastly, the influence of friction on the electrical energy consumption of the production process [38]. Seki investigated the influence of friction on the deformation processing of wood, regarding the contact pressure [39], the moisture content [40] and the surface condition of the wood and the tools [41].

Structural design of connections and constructions

Early experiments to determine the coefficient of friction in timber-timber connections were performed by Gaber [8] and date back to 1940. Practical needs were the driving force for the tests, as the determined friction coefficients were to be used in calculations of timber trusses. Gaber has also experimented with increasing the friction in the shear plane by adding sand grains between the timber parts. However, the friction coefficient decreased if too much sand or grain sizes that were too large were added. According to Gaber, it is better to make connections without sand because the risk of failure is too great. Around the same time, Johansen [15] performed friction tests to back his *Theory of Timber Connections*.

Möhler & Maier [9] investigated the static coefficient of friction for wood on wood, wood on steel, and wood on concrete for the use of timber falsework (Figure 2.4). They determined a dependency on the moisture content and the contact pressure (for low pressures only). Like Gaber, they experimented with sand to increase the friction coefficient between the planks of the timber arch rib. At low contact pressure, the sand grains were not pressed in, and the timber rolled over them, while at high pressure, the sand grains got stuck in the wood. This also resulted in the recommendation to refrain from using sand, as the certainty of success is too low. Further investigations by Möhler & Herröder [42, 43] of the coefficient of friction for falsework constructions showed almost no difference in the friction coefficient between parallel and perpendicular fibre direction. Tests with end grain, however, resulted in significantly higher friction coefficients. The coefficients of friction determined in these tests are exactly reproduced in Appendix B of the German version of EN 12812 [110] *Falsework – Performance requirements and general design* (formerly DIN 4421). Gorst et al. [44] also performed tests on 260 combinations to determine the friction coefficient for wood on engineered wood products used in falsework.

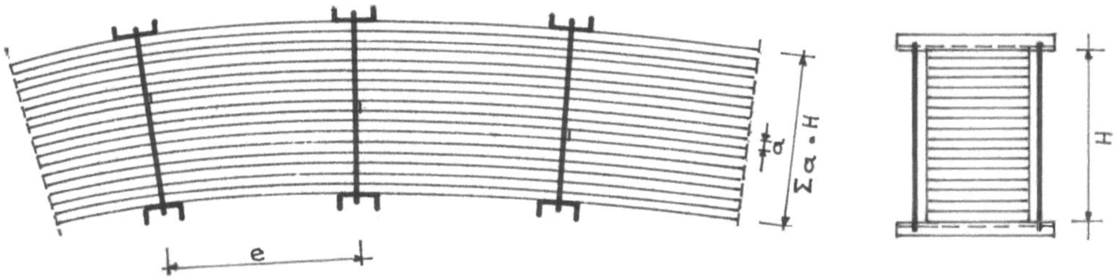

Figure 2.4: Setup of an arch rib according to Cruciani [9].

In traditional carpentry joints the load is often transferred by direct contact of the two adjoining timber members. Much research has been performed in the last two decades to account for the resulting friction in modelling such joints. Bejo et al. [45] investigated the influence of the contact pressure on the friction coefficient for timber and timber. They concluded that a significant influence of the contact pressure led to lower friction coefficients with higher contact pressure. Further tests regarding friction in carpentry joints were performed by Crespo et al. [46] (see also [47]) and Park et al. [48], investigating traditional Korean timber structures.

For the structural design of light-frame timber shear walls, Steiger et al. [49] determined the coefficient of friction for wood-based panels on wood, and Claus et al. [50] determined the kinetic coefficient of friction for wood on wood. For the earthquake design of timber floor connections in masonry buildings, Almeida et al. [51] determined the kinetic coefficient of friction for wood on wood (see also [52]). Meng et al. [53] investigated the coefficient of friction for wood on different wood-based panels and wood on steel. The obtained data was used to design timber light-frame shear walls and mechanical timber joints [54].

For the design of mechanical timber joints with dowel type fasteners, Sjödin et al. [55] estimated the friction coefficient for steel on wood by evaluating test data of joint tests. They showed that higher load-carrying capacities are possible if dowels with a rough surface are used (Figure 2.5). The same observation was made by Dorn [56], who also performed tests with dowels with different surface roughness. Blaß & Steige [57] also estimated the coefficient of friction for wood on wood using test data of timber joint tests. Based on these observations, Dorn et al. [58] performed fundamental tests to determine the coefficient of friction for steel on laminated veneer lumber (LVL) and timber, varying between many parameters (see also [59]). To determine the coefficient of friction for a steel dowel on wood, Rodd [60] performed some tests using semi-circular steel blocks and varying the contact pressure. Further tests were performed at KIT Timber Structures and Building Construction as part of the European transnational project *hardwood_joint* [21] to experimentally investigate the friction between beech LVL and steel dowel, varying the fibre direction (see also

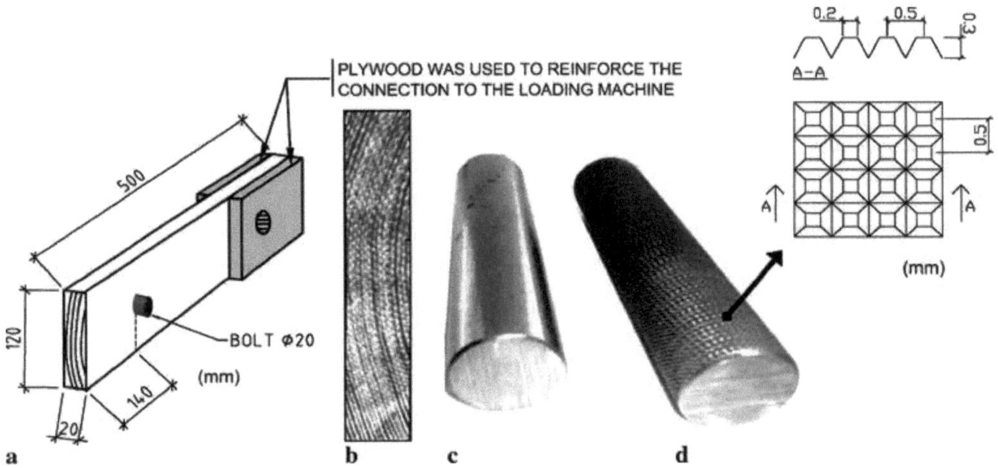

Figure 2.5: Investigated dowels with smooth (c) and rough surface (d) [55].

For the invention of a new shear connector, Schmidt [61] determined both the static and kinetic friction coefficient for cross-laminated timber (CLT) on steel, varying the contact pressure and the sliding speed (see also [62]). To improve existing system connectors by increasing the friction in the shear plane between timber and connector, tests were performed at KIT Timber Structures and Building Construction [63] to determine the coefficient of friction for wood on aluminium. Tests were performed with untreated connectors and connectors coated with an anti-slip coating (Figure 2.6a). In a recently completed research project [64] hybrid timber trusses were investigated. Hybridization was done through the use of hardwood LVL and softwood glulam. To increase the stiffness of the connection of the web and chord, the web's surface was modified to increase the coefficient of friction of hardwood on softwood (Figure 2.6b).

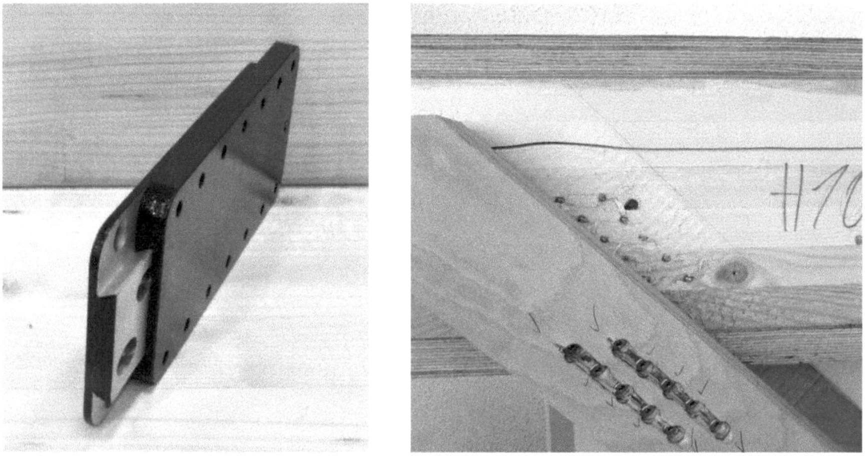

(a) (b)

Figure 2.6: (a) Investigated system connectors with coated surface [63] and (b) joint of hybrid truss with increased friction in the shear plane [64].

Test setup

During the literature review, different test setups were encountered. The most common test setups can be seen in Figure 2.7. Setup (a) with the inclined plane is very easy to use. The normal force is applied with weights, and the plane is inclined until the specimen starts sliding. However, due to the constantly changing contact pressure depending on the angle, the inclined plane is unsuitable for determining friction coefficients reliably [12, 30]. In the evaluated literature, the horizontal plane (b) was the most used test setup because of its simple design. This setup allows the evaluation of all parameters (to some extent). The contact pressure is applied by weights, which is a limiting factor. Test setup (e) is like the horizontal plane, with the only difference being that a vertical cylinder is used to apply the contact pressure. It, therefore, is suitable for investigating the influence of high contact pressure. The rotating plane in setup (c) was mainly used for tests with wood sliding on steel. With this setup, it is possible to study the influence of very high sliding speeds. A practical setup for higher contact pressure without using a second cylinder is setup (d) with pre-stressed rods. However, the specimen size is much larger than for the other tests, and the time to assemble the specimen takes longer. Also, decreasing contact pressures are possible due to losses of the pre-stress loads. The last setup (f) for shear tests with inclined screws is not primarily used for the determination of the coefficient of friction. However, the coefficient can be evaluated from test data [57]. This setup might, therefore, give the most realistic values for the friction coefficient for the corresponding application case.

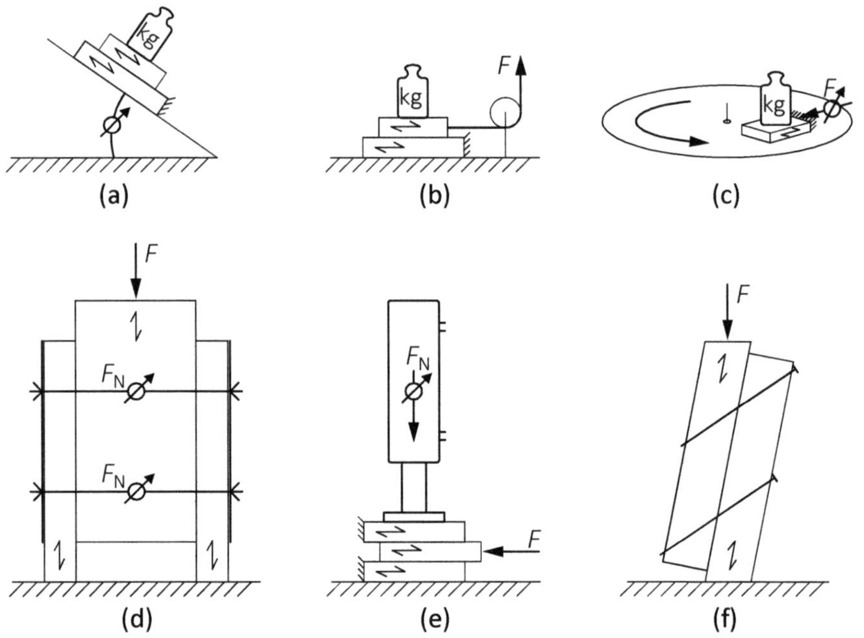

Figure 2.7: Test setups to determine the coefficient of friction: (a) inclined plane; (b) horizontal plane; (c) rotating plane; (d) pre-stressed rods; (e) two hydraulic cylinders; (f) shear tests with inclined screws.

Friction coefficients from literature

Table 2.1 lists static and kinetic friction coefficients for softwood on softwood and hardwood on hardwood. Table 2.2 lists friction coefficients for softwood and hardwood on steel or aluminium. Table 2.3 lists friction coefficients for engineered wood products.

The given friction coefficients in the tables are for a moisture content of $u < 20\%$. For the sake of clarity (and as the evaluation of data suggests), it was not distinguished between different grain directions of the wood, and different surfaces of the steel or aluminium in

Tables 2.1–2.3. The number of values taken from the literature is also given. This number does not necessarily correspond to the tests carried out. For example, in [12] a total of 9000 tests are mentioned, but only ten values can be taken from the two-page article.

Table 2.1: Static and kinetic friction coefficients from literature for tests with wood on wood. Values of μ_s and μ_k only for $u < 20\%$.

Publication	n	μ_s	μ_k
SOFTWOOD		ON SOFTWOOD	
Aira et al. [47]	9	0.05–0.38	0.30–0.25
Almeida et al. [51]	6	0.31–0.69	0.27–0.65
Blaß & Steige [57]	79	0.02–1.03	-
Claus et al. [50]	19	0.14–0.89	0.11–0.56
Crespo et al. [46]	10	0.39–0.53	0.21–0.45
Gaber [8]	41	0.17–0.92	-
Gorst et al. [44]	39	0.30–0.80	-
Gressel & Redecker [65] in [66]	1	0.34	0.25
Johansen [15]	2	0.40–1.40	-
Koch [67]	6	0.37–0.66	-
McKenzie & Karpovich [31]	3	0.45–0.60	-
Möhler & Maier [9]	94	0.21–0.98	0.25–0.70
Möhler & Herröder [42]	44	0.29–1.37	0.14–0.78
Murase [29]	8	0.60–0.68	0.56–0.60
Park et al. [48]	20	0.44–0.74	0.29–0.60
Stošié [12]	5	0.30–0.49	-
Xu et al. [68]	15	0.40–0.60	0.25–0.61
	sum = 401	mean = 0.49	mean = 0.40
HARDWOOD		ON HARDWOOD	
Almeida et al. [51]	3	0.59–0.61	0.56–0.57
Gorst et al. [44]	12	0.40–0.60	-
Gressel & Redecker [65] in [66]	3	0.28–0.46	0.22–0.35
Murase [29]	8	0.53–0.60	0.50–0.59
Stošié [12]	3	0.30–0.31	-
Xu et al. [68]	30	0.30–0.78	0.14-0.61
	sum = 59	mean = 0.48	mean = 0.38

n = number of values taken from respective literature

Table 2.2: Static and kinetic friction coefficients from literature for tests with wood on steel. Values of μ_s and μ_k only for u < 20%.

Publication	n	μ_s	μ_k
SOFTWOOD		ON STEEL / ALUMINIUM	
Atack & Tabor [11]	1	-	0.60 [1)]
Dorn et al. [58] & [59]	102	0.10–0.90	-
KIT Timber Structures [21]	4 [2)]	0.24–0.38	0.17–0.27
Gorst et al. [44]	102	0.30–0.70	-
Guan et al. [36]	18	-	0.16–0.60
Kuwamura [30]	106	0.10–0.73	-
Lemoine et al. [32]	48	-	0.10–0.57
McKenzie & Karpovich [31]	23	0.11–0.65	0.15–0.40
Meng et al. [53]	24	0.22–0.36	0.17–0.28
Möhler & Herröder [42]	20	0.48–1.17	0.35–0.85
Murase [29]	24	0.12–0.22	0.12–0.20
Rodd [60]	120 [2)]	-	0.30–0.49
Schmidt [61]	72	0.28–0.47	0.19–0.46
Seki et al. [41]	21	0.12–0.39	-
	sum = 685	mean = 0.32	mean = 0.24
HARDWOOD		ON STEEL / ALUMINIUM	
Gorst et al. [44]	63	0.30–0.70	-
Guan et al. [36]	6	-	0.20–0.30
KIT Timber Structures [21]	8 [2)]	0.23–0.31	0.14–0.23
McKenzie & Karpovich [31]	50	0.08–0.64	-
McLaren & Tabor [28]	4	0.10–0.50	-
Price & Manwiller [69] in [70]	132	-	0.11–0.13
	sum = 131	mean = 0.39	mean = 0.20

n = number of values taken from respective literature

[1)] u < 30%

[2)] tests with steel dowels

Table 2.3: Static and kinetic friction coefficients from literature for tests with engineered wood products.

Publication	*n*	μ_s	μ_k
ENGINEERED WOOD	ON SOFTWOOD / HARDWOOD		
Bejo et al. [45]	16	0.33–0.84	0.20–0.52
Gorst et al. [44]	153	0.10–0.60	-
KIT Timber Structures [21]	20	0.30–0.48	0.24–0.48
Meng et al. [53]	192	0.23–0.42	0.15–0.36
Gressel & Redecker [65] in [66]	17	0.12–0.59	0.11–0.35
Steiger et al. [49]	2	0.24–0.27	0.19–0.21
	sum = 400	mean = 0.30	mean = 0.24
ENGINEERED WOOD	ON STEEL / ALUMINIUM		
Dorn et al. [58] & [59]	60	0.12–0.63	-
Gorst et al. [44]	168	0.10–0.70	-
KIT Timber Structures [21]	36 1)	0.10–0.40	0.07–0.36
	sum = 264	mean = 0.23	mean = 0.18

1) tests with steel dowels

2.3 Evaluation of the results from literature

Moisture content

The moisture content u is the only parameter given for almost all tests. The moisture content of the wood specimens has the most distinct influence on the friction coefficient (see Figure 2.8 and Figure 2.9). Especially for the tests with wood on wood, a moisture content greater than 20% leads to a notable increase of the friction coefficient. This is independent of softwood, hardwood or engineered wood products. The tests with steel/aluminium on wood are primarily independent of the moisture content of the wood for $u < 20\%$, and slightly increase for $u > 20\%$. For most tests, however, no explicit value for the moisture content was given; instead, it was a range of the moisture content. It cannot be evaluated if the friction coefficient increases suddenly (as the diagrams might suggest) or steadily with increasing moisture content. For timber engineering, a moisture content > 20% should be avoided. Therefore, the evaluation of further parameters only includes test results with a moisture content < 20%. Tests with a moisture content of > 20% are no longer considered, as they are deemed irrelevant for practical use in timber engineering.

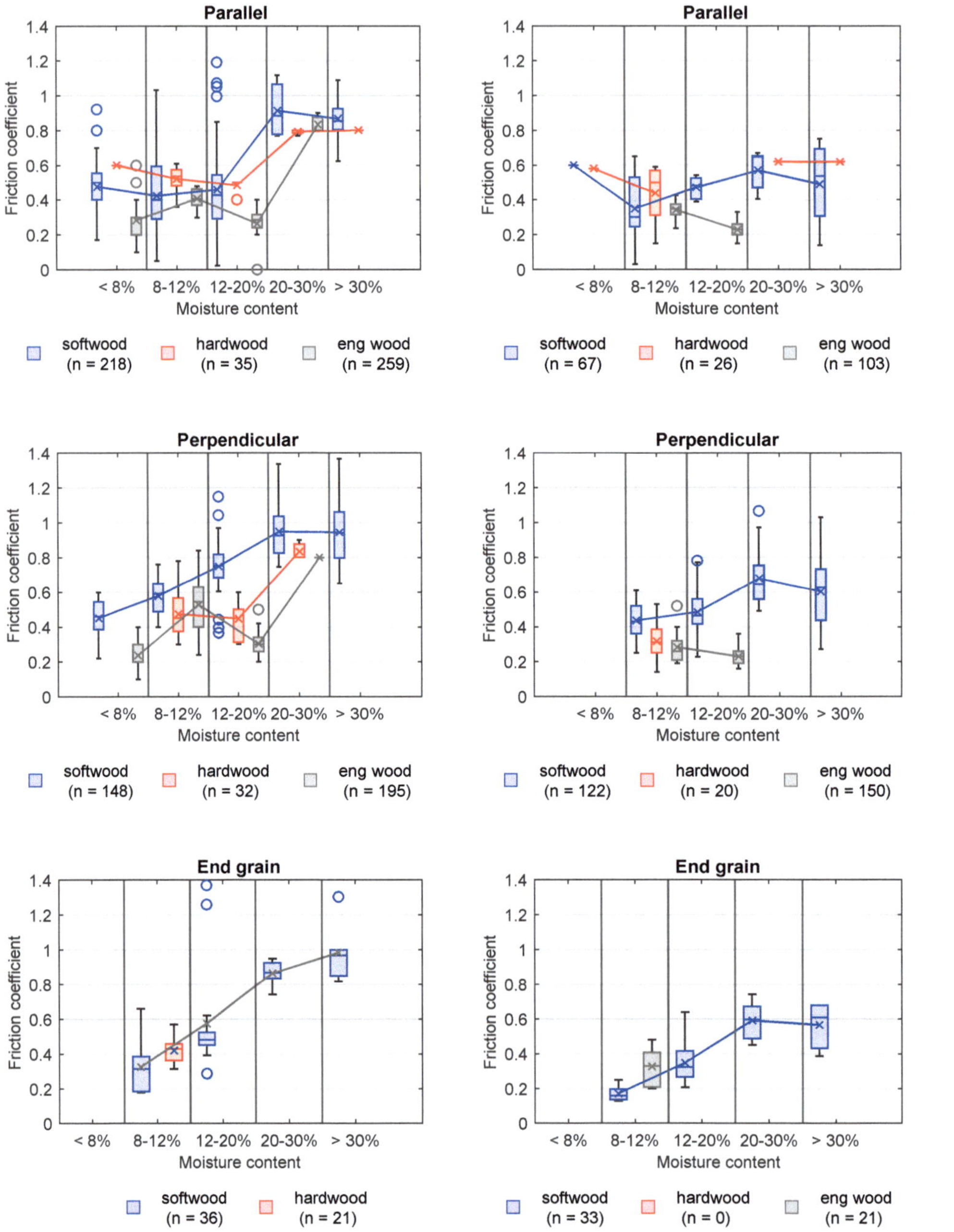

Figure 2.8: Influence of moisture content for wood on wood. Parallel, perpendicular and end grain. Static (left) and kinetic (right) coefficient of friction.

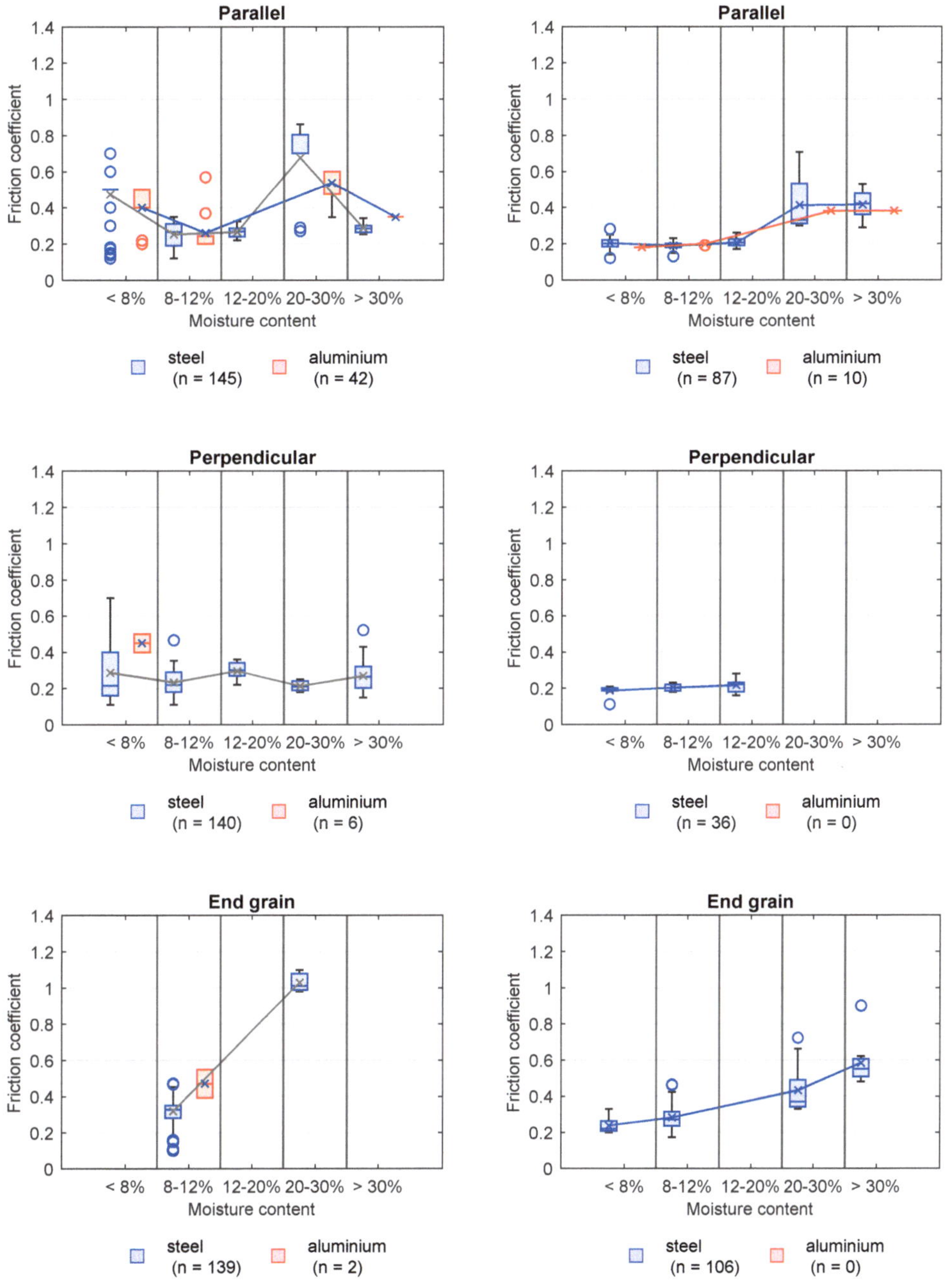

Figure 2.9: Influence of moisture content for steel on wood. Parallel, perpendicular and end grain. Static (left) and kinetic (right) coefficient of friction.

Grain direction

In many tests, the orientation of the fibres regarding the sliding direction is varied, with the different orientations being parallel (grain direction of the specimens parallel to each other), perpendicular (one specimen rotated by 90°) and end grain (end grain sliding on end grain). In some publications, further angles are examined, but for clarity, only the three main directions, parallel, perpendicular and end grain, are considered. Figure 2.10 clearly shows no differences between tests parallel and perpendicular to the grain. For the tests with end grain, there is only a slight increase in the friction coefficient. Therefore, the grain direction has close to no influence on the friction coefficient. The same assumption can be made for tests with steel/aluminium on wood. Therefore, the following evaluation no longer distinguishes between parallel and perpendicular grain directions. For clarity, tests with end grain on end grain are no longer considered.

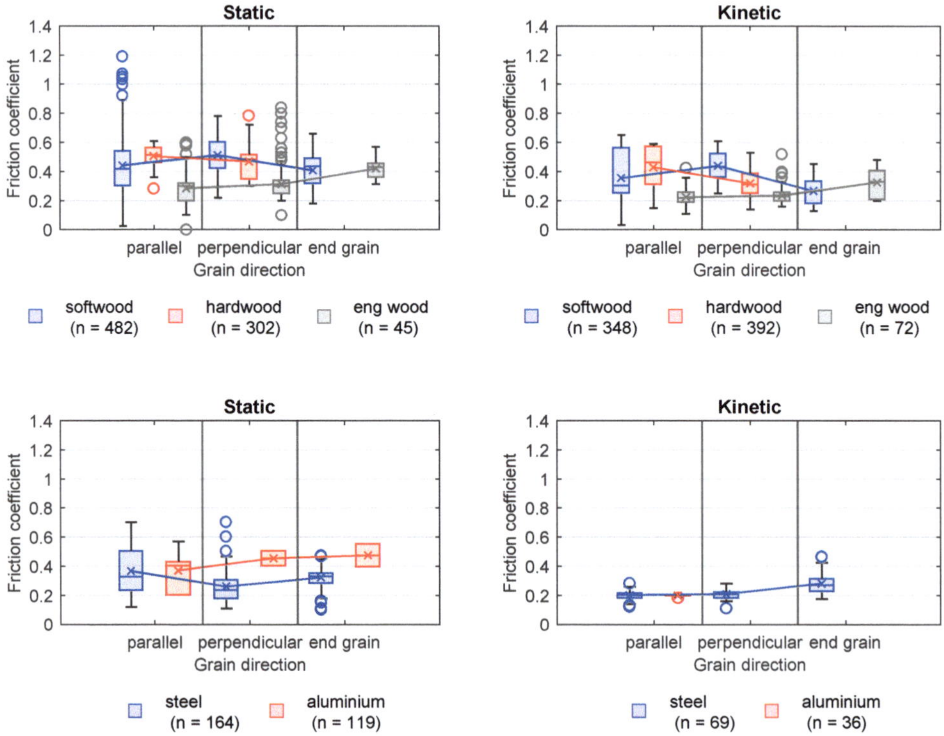

Figure 2.10: Influence of grain direction for wood on wood and steel on wood. Static and kinetic COF.

Contact pressure

Figure 2.11 shows the influence of the contact pressure on the friction coefficient. The recorded pressures range from 0.0001 to 2.5 N/mm^2 for tests with wood on wood and 0.0069 to 35 N/mm^2 for tests with steel or aluminium on wood. No clear influence of the contact pressure on the friction coefficient is visible for low contact pressures due to the high scatter of the values. The friction coefficient seems to increase with increasing pressure, especially for tests with hardwood on hardwood. However, the R^2-value is as low as 0.22. Interestingly, the coefficient of friction for tests with steel/aluminium on wood seems to decrease with increasing pressure. But again, the R^2-values are very low.

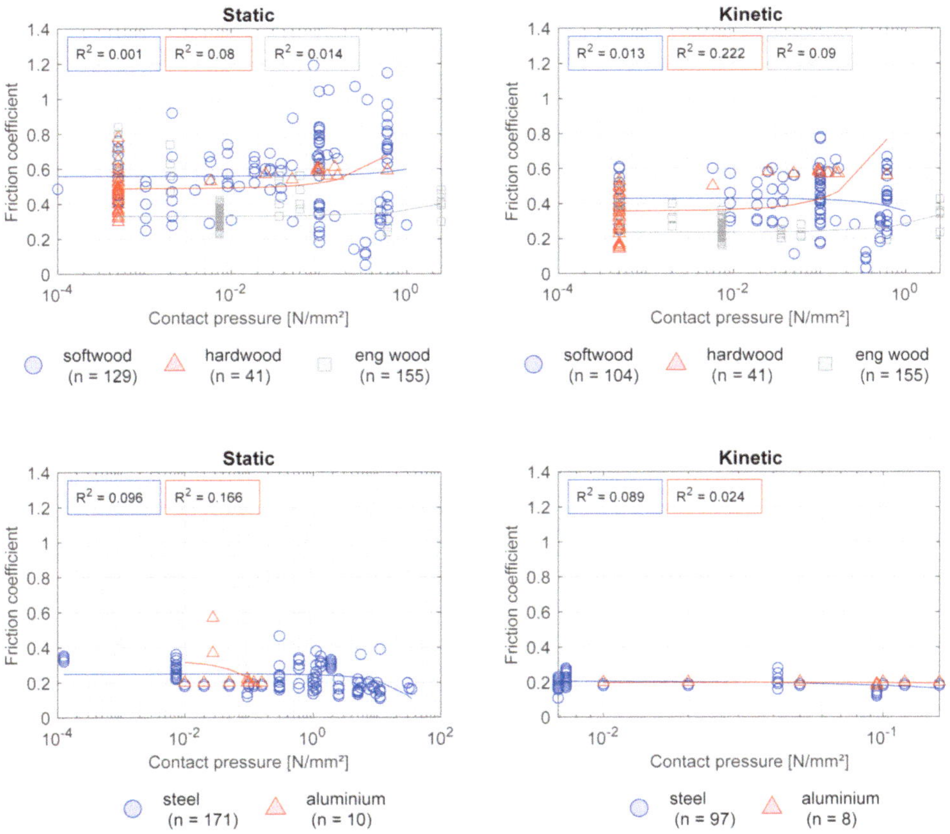

Figure 2.11: Influence of contact pressure for wood on wood and steel on wood. Static and kinetic COF.

Sliding speed

The recorded sliding speeds range from 1 to 3 300 mm/min for tests with wood on wood and 1 to 2 640 000 mm/min (= 44 m/s) for tests with steel/aluminium on wood. Figure 2.12 shows no influence of the sliding speed on the friction coefficient for lower speeds. Only for tests with hardwood does there seem to be a reduction in the friction coefficient with higher sliding speeds. Again, the R^2-values are very low. The R^2-value indicates how good the trend line predicts the actual results. To emphasize the difficulty (and inaccuracy) of predicting an influence of the single parameters, the R^2-value is deemed to be a good indicator. For the tests with steel on wood, the friction coefficient slightly increases with increasing sliding speed (R^2 = 0.13–0.30).

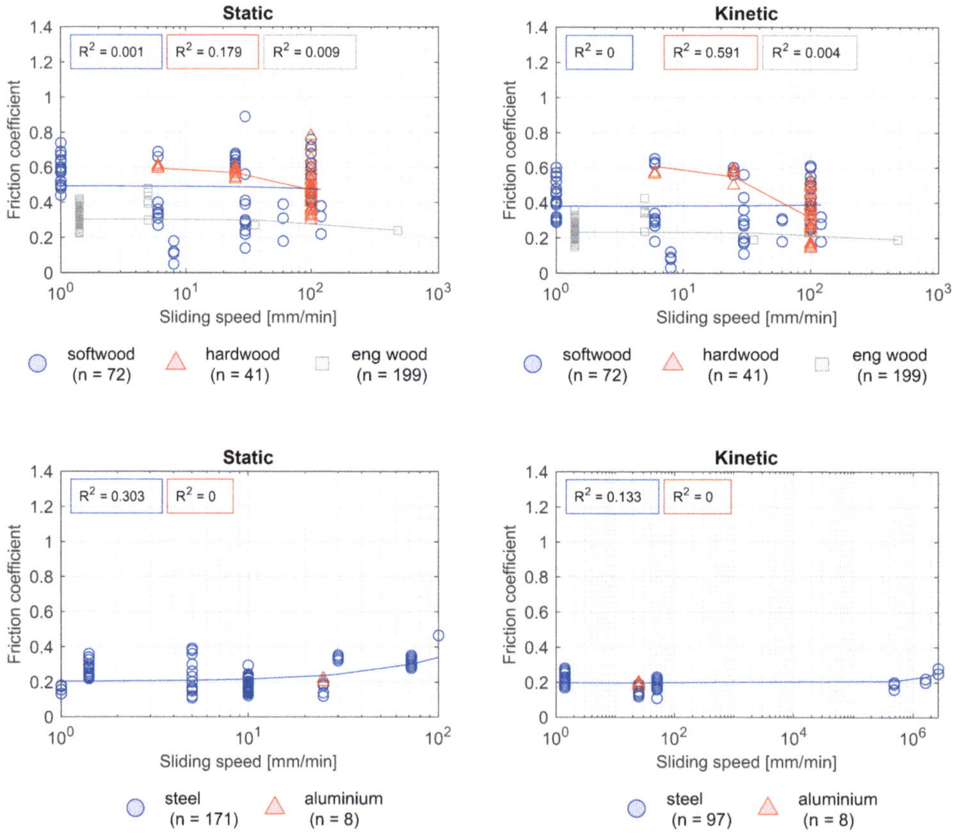

Figure 2.12: Influence of sliding speed for wood on wood and steel on wood. Static and kinetic COF.

Density

Figure 2.13 shows no correlation between the friction coefficient and the density of the wood specimens. With the only exception being tests for kinetic friction with hardwood on hardwood (R^2 = 0.41). However, the scatter of results is significant. The scatter decreases for the tests with steel/aluminium on wood. Here, the friction coefficient slightly decreases with increasing density.

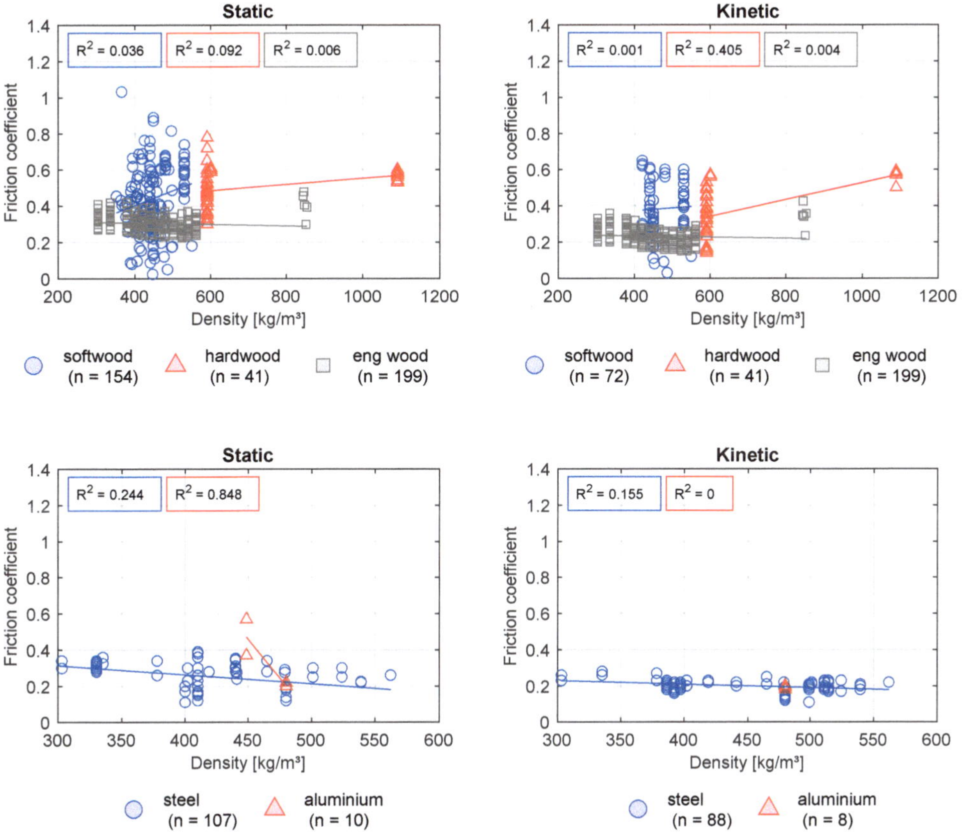

Figure 2.13: Influence of bulk density for wood on wood and steel on wood. Static and kinetic COF.

Surface condition

The surface roughness of the tested specimens influenced the friction coefficient. Figure 2.14 shows a box plot diagram with three different surfaces for wood on wood (top) and four different surfaces for steel on wood (bottom). Treated surfaces were, for example, falsework panels. The classification "normal steel" was used when no other surface quality was explicitly mentioned. The static friction coefficient increases with increasing surface roughness. The influence of the surface roughness on the kinetic friction coefficient is not as distinctive, except for tests with hardwood.

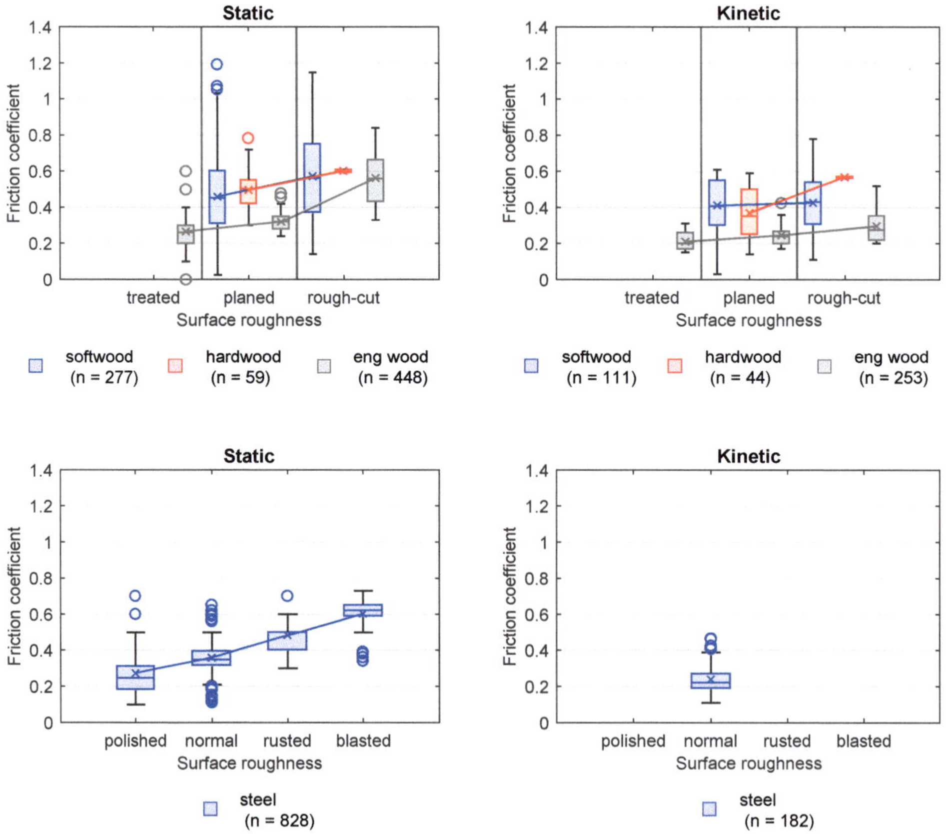

Figure 2.14: Influence of surface condition for wood on wood. Static and kinetic COF.

Scatter

As is evident now, the friction coefficient values vary largely for the same tested materials and parameters. To highlight the scatter, Figure 2.15 shows an exemplary plot for the friction coefficient for softwood on softwood, hardwood on hardwood, engineered wood on either softwood or hardwood, and steel or aluminium on softwood. The sliding direction for all tests was parallel to the grain of the wood specimens. The moisture content was $\leq 20\%$, and the wood surface was planed. The scatter for static tests with engineered wood products, steel, and aluminium is smaller than for tests with softwood and hardwood. Also, the scatter of kinetic coefficients is smaller than for static coefficients. The coefficient of variation of the static coefficient of friction for softwood is around 45%, for hardwood 20% and for engineered wood products 20%. The coefficient of variation for steel is 31% and for aluminium 30%.

The data scattering between the diagrams is different because the database was filtered for each parameter. Not all parameters were given for all results; therefore, each dataset for each parameter is unique. Each diagram presented here in the evaluation is, therefore, based on a separate data set. This explains the different extreme values of the coefficient of friction in the various diagrams and, thus, the different deviations.

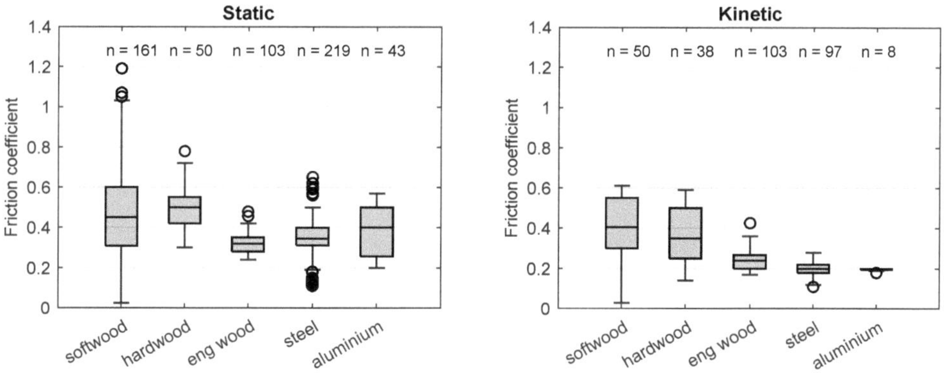

Figure 2.15: Scatter of all materials. Static and kinetic COF.

Test setup

Figure 2.16 shows the influence of the test setup on the friction coefficient. Again, only tests with a moisture content of less than 20% were considered. Indeed, higher test results are reported when using test setup (a) compared to test setup (b), but the highest friction coefficients were still reported when using test setup (e). Therefore, it is not easy to make a clear statement about the influence of the test setup.

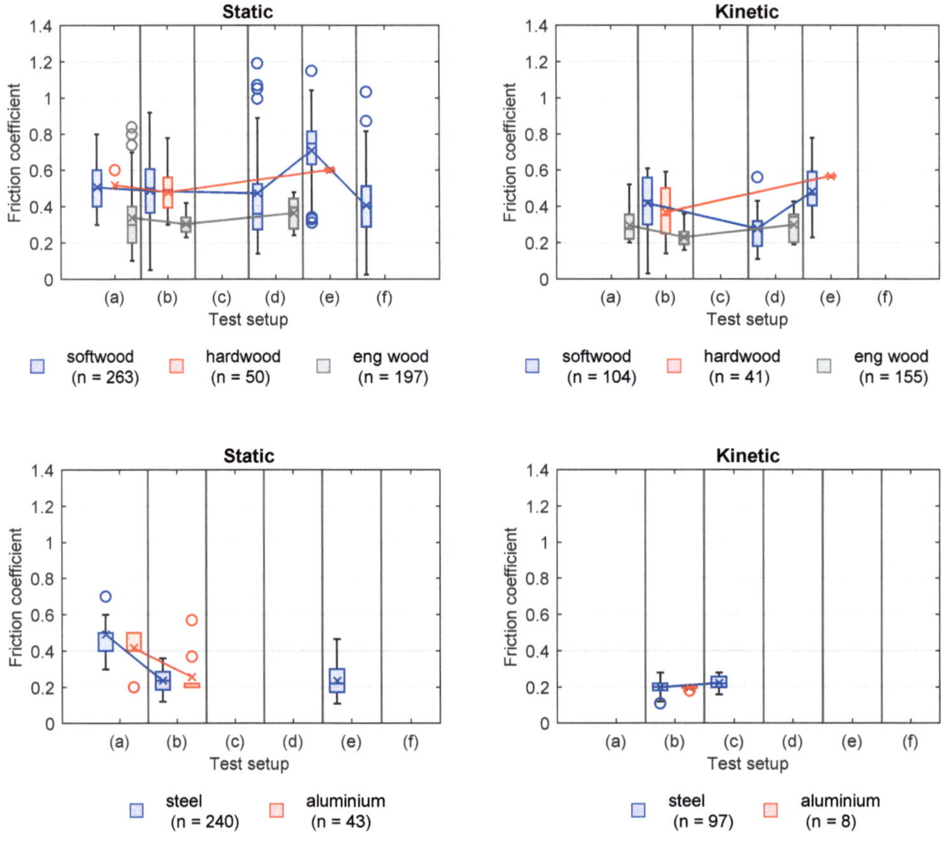

Figure 2.16: Influence of test setup for wood on wood and steel on wood. Static and kinetic COF.

2.4 Comparison to values given in standards

EN 1995-1-1

According to the design model of connections with dowel-type fasteners in Eurocode 5 [114], a friction coefficient of 0.25 shall be used to design both timber-to-timber and steel-to-timber connections. No unique source found in the literature shows how the value 0.25 was determined. However, if all values from the reviewed literature are considered (only values for tests with moisture content < 20%, but no further distinction of wood species, surface type, etc.), a mean static coefficient of friction of 0.39 and 0.36 can be calculated for tests with wood on wood and steel/aluminium on wood, respectively. Again, assuming a log-normal distribution, 5%-quantile values can be determined. With a global coefficient of variation of $COV_g = 0.16$ and a global $k_{s(n)}$-value of 1.76 (see Chapter 4, Section 4.5 for explanation and determination of COV_g and $k_{s(n)}$) 5%-quantiles of 0.27 and 0.24 can be calculated for tests with wood on wood and steel/aluminium on wood, respectively. This adequately confirms the value of 0.25 in Eurocode 5 as the (characteristic) value of the friction coefficient of wood or steel/aluminium on wood, with no differentiation of wood species and grain direction.

EN 1995-2

Part 2 of Eurocode 5 [115] contains general principles for the design and construction of the main structural components of bridges. Table 6.1 gives the design values of friction coefficients for stress-laminated deck plates. Again, it is unclear where these values come from and how they were determined. To compare the result from the literature with the results from the standard, firstly, characteristic values were determined with a global coefficient of variation of $COV_g = 0.16$ and a global $k_{s(n)}$-value of 1.76. Secondly, these 5%-quantiles were then converted to design values with $\gamma_M = 1.3$ and $k_{mod} = 1.0$ and are given in parentheses in Table 2.4. The values generally agree well, considering that the correct selection of k_{mod} brings the calculated values closer to the table values. Interestingly, Part 2 of Eurocode 5 specifies the contact pressure in the friction surface: the initial contact pressure should be at least 1.0 N/mm^2 and should not fall below 0.35 N/mm^2 during the service life.

Table 2.4: Friction coefficients (design values) according to Table 6.1 in EN 1995-2 and design values calculated from literature (in parentheses).

Surface roughness	Perpendicular to grain		Parallel to grain	
	$u \leq 12\%$	$u \geq 16\%$	$u \leq 12\%$	$u \geq 16\%$
Timber sawn / sawn	0.30 (0.40)	0.45 (0.50)	0.23 (0.30)	0.35 (0.45)
Timber planed / planed	0.20 (0.15)	0.40 (0.50)	0.17 (0.20)	0.30 (0.35)
Timber sawn / planed	0.30	0.45	0.23	0.35
Timber / concrete	0.40	0.40	0.40	0.40

EN 12812 (German version)

As mentioned before, friction coefficients for wood are also given in Table B.1 in Appendix B of the German version of EN 12812; see Table 2.5. The values were determined in [42] and are given as mean values as obtained in the tests. Additionally, Appendix B states that "the characteristic coefficients of friction may be derived from the results of other research projects" [110].

Table 2.5: Static friction coefficients given in Table B.1 in EN 12812 (mean values).

Material combination	Coefficient of friction μ	
	Maximum	Minimum
Timber/timber – parallel or perpendicular	1.0	0.4
Timber/timber – end grain	1.0	0.6
Timber/steel	1.2	0.5
Timber/concrete	1.0	0.8
Steel/steel	0.8	0.2
Steel/concrete	0.4	0.3
Steel/mortar	1.0	0.5
Concrete/concrete	1.0	0.5

2.5 Parameter study based on own experiments

The literature review showed that comparing the results of different research projects is difficult. Not all parameters are recorded for all tests, and the information is sometimes inaccurate. Therefore, an experimental parameter study was performed to validate the findings from the literature. The varied parameters were the contact pressure, the sliding speed, and the grain direction. Additionally, the density was recorded for all specimens and two different test setups were used. Test were performed with densified veneer wood (DVW) on softwood. The used materials were softwood glulam GL 24h, which was conditioned at 20°C and 65% r.H., and densified veneer wood (DVW). For more information on DVW, refer to [3]. The surface of the DVW specimens for the parameter study was untreated and smooth. It was distinguished between three different grain directions of the softwood glulam, i.e. face grain perpendicular (Figure 2.20a), face grain parallel (Figure 2.20b), and end grain (Figure 2.20c).

2.5.1 Test setup and execution

The experimental setup for the tests to determine the coefficient of friction of DVW on softwood was adopted from Schmidt [61], see Figure 2.17. The experimental frame consisted of two thick metal plates held together with four threaded rods. On one side, a spindle was welded to the metal plate. A threaded rod with a fine thread was turned through this spindle. The force F_n perpendicular to the friction surface was applied via this threaded rod. Due to the fine thread, F_n could be precisely set. During the tests, the force F_N was measured continuously with a load cell placed directly behind the threaded rod. To distribute F_n evenly over the friction surface, a calotte was placed between the load cell and the friction surface. In addition, the softwood pieces were placed on metal blocks so that the force F_n acted centrally on the friction surface in the axis of gravity of the threaded rod. The frictional force F_f parallel to the friction surface was applied to the test specimens via a universal testing machine. The entire test procedure was displacement-controlled up to a displacement of 15 mm (of the machine head). The sliding speed was varied as part of the parameter study. In total, 58 tests were performed with this setup. The tests were performed by Schiebel [71] as part of his Bachelor's Thesis.

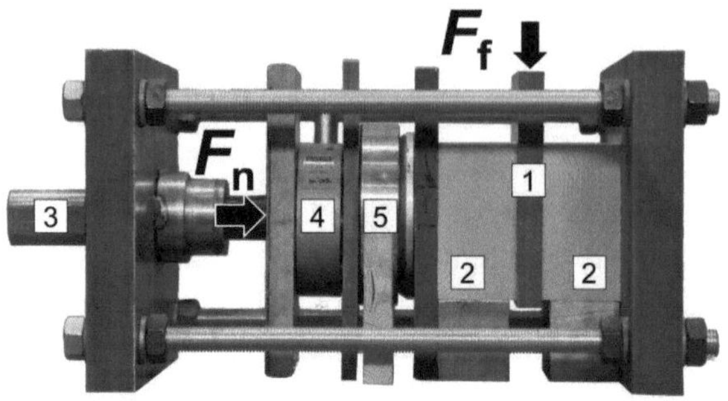

1 test specimen with modified surface
2 softwood side members
3 spindle to apply F_n
4 load cell for measuring F_n
5 calotte for load distribution

Figure 2.17: Test setup 1 for friction tests [71].

The disadvantage of the setup with the threaded rod was that the force F_n decreased during the testing. With the threaded rod and the spindle, it was impossible to adjust the normal force to the desired value (and keep it) once the test ran. This disadvantage was not deemed problematic to evaluate the friction coefficient, as the force was measured continuously, and the friction coefficient was always calculated as the ratio of the actual normal force to the frictional force (see for example Figure 4.2b). Nevertheless, the first setup was upgraded later in the research project. The basic setup and all the steelwork remained the same. However, a hydraulic cylinder replaced the spindle with the threaded rod (see Figure 2.18). The hydraulic cylinder was load-controlled, and it was now possible to set and keep the normal force precisely at the desired value. In total, 141 tests were performed with this setup. The tests were performed by Albicker [72] as part of his Bachelor's Thesis.

1 test specimen with modified surface
2 softwood side members
3 hydraulic actuator to apply F_n
4 load cell for measuring F_n
5 calotte for load distribution

Figure 2.18: Test setup 2 for friction tests [72].

Exemplary load-displacement curves for tests performed with setup 1 and setup 2 can be seen in Figure 2.19. The load-displacement curve of the normal force F_n of setup 1 (blue line in Figure 2.19a) showed an initial increase in load when the specimen was loaded parallel to the friction surface. The normal force then decreased continuously once the maximum frictional force F_f (red line) was reached. In contrast, the load-displacement curve recorded with setup 2 (blue line in Figure 2.19b) was constant during the whole test and neither increased nor decreased. In both diagrams, the curve of the friction coefficient over machine displacement is given (grey lines). It clearly shows that the constant decrease of the normal force in setup 1 has no visible influence on the friction coefficient's size and behaviour.

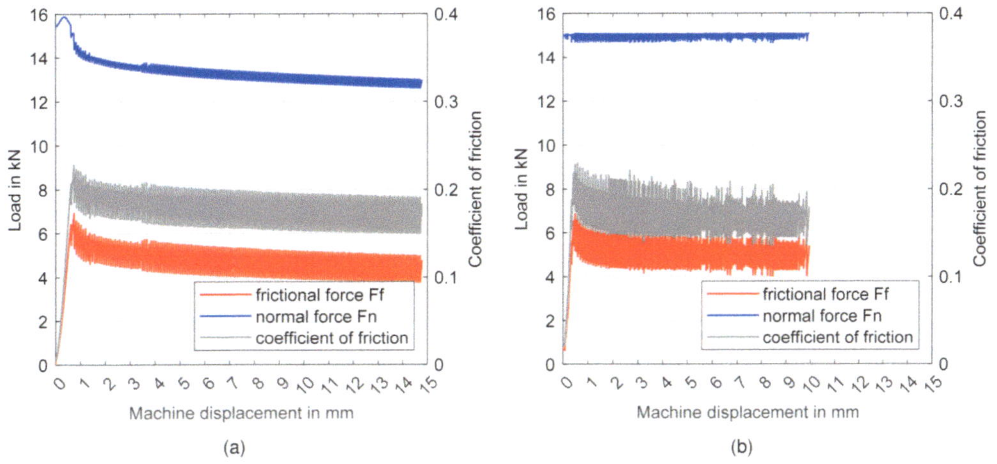

Figure 2.19: Load-displacement curves for test (DVW with untreated surface on softwood face grain perpendicular) performed with setup 1 (a) and setup 2 (b).

During the tests, a distinction was made between the softwood's face grain and end grain, to consider different installation situations of the connectors. Contact with the face grain occurs on the main beam/header, whereas contact with the end grain occurs on the secondary beam/joist. Furthermore, a distinction was made between the grain direction of the wood specimens parallel to or perpendicular to the loading direction. The different test configurations can be seen in Figure 2.20. The different sizes of the friction surface are also evident, but this was considered via the force F_n. For the tests with face grain, the friction surface was 100x100 mm^2; for the tests with end grain, the friction surface was 50x100 mm^2. The contact pressure in the friction surface was varied as part of the parameter study.

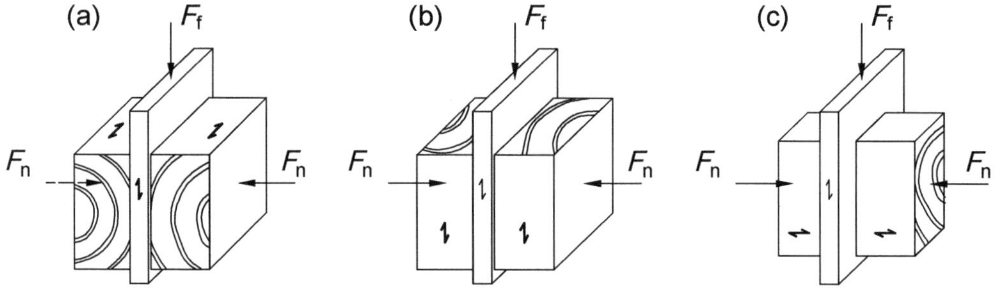

Figure 2.20: Investigated grain directions of the wood specimens concerning the sliding direction: (a) face grain perpendicular; (b) face grain parallel; (c) end grain.

The static coefficient of friction was calculated according to Equation 2.4. The force F_f parallel to the contact surface was divided equally between the two friction surfaces. Three typical behaviours of surfaces are shown in Figure 2.2, and $F_{f,static}$ was chosen accordingly. In cases where stick-slip occurred, the load $F_{f,static}$ was chosen as the maximum force at the very first peak, according to American standard ASTM G115 [105].

The kinetic coefficient of friction was calculated with the force $F_{f,kinetic}$ corresponding to the horizontal part of the load-displacement curve. As that part was not always explicit (slope of the curve still increasing/decreasing or any other form of non-linear behaviour), the kinetic coefficient of friction was evaluated as the mean value between a displacement of 10 to 15 mm. The range of 10–15 mm was chosen as this was the range where most curves ran horizontally. With so many tests and sometimes quite different curves in one series, it was impossible to evaluate the friction coefficients automatically with a MATLAB®-script [119]. Therefore, some friction coefficients had to be evaluated manually, contrary to the approaches mentioned.

$$\mu = \frac{F_f}{2 \cdot F_n} \tag{2.4}$$

with
μ coefficient of friction
F_f frictional force
F_n normal force

2.5.2 Results

Influence of grain direction

According to Figure 2.20, the three main grain directions were investigated, i.e. sliding direction parallel to the grain, perpendicular to the grain, and tests with end grain. The results in Figure 2.21a show no difference for tests with softwood parallel and perpendicular to the grain but slightly lower results for tests with end grain. An analysis of variance (one-way ANOVA) was performed to check for differences in the various test series (for a more detailed explanation, see Chapter 4, Section 4.4). The ANOVA determined that there is a statistically significant difference in friction coefficients ($F(2, 196) = [14.72]$ and $p < 0.0001$). The subsequent Tukey's HSD test for multiple comparisons showed that the mean value of friction coefficients significantly differed between tests with face grain and end grain. There was, however, no statistically significant difference in friction coefficients between the sliding direction parallel to the grain and perpendicular to the grain. This confirms the evaluation of the results from the literature.

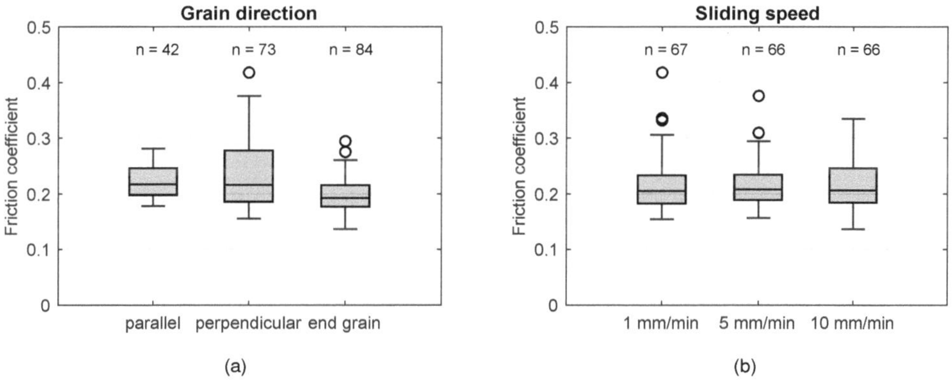

Figure 2.21: Influence of (a) grain direction and (b) sliding speed on coefficient of friction.

Influence of sliding speed

Three different sliding speeds were investigated, i.e. 1, 5 and 10 mm/min. The mean friction coefficients of all three sliding speeds are similar, as shown in Figure 2.21b. This is also confirmed with a one-way ANOVA. There is no statistically significant difference in friction coefficients between different sliding speeds in tests with face grain ($F(2, 112) = [0.00]$ and $p = 0.9989$) and no statistically significant difference in tests with end grain

(F(2, 81) = [0.17] and p = 0.8476). Again, this confirms the evaluation of the results from the literature.

Influence of contact pressure

Five contact pressures were investigated: 0.25 (end grain only), 0.5, 1.0, 2.5 and 6.0 N/mm^2. Figure 2.22a shows the results for all three grain directions. Contrary to the literature review, a slight decrease in the friction coefficient was observed with increased contact pressure. However, the decrease is very shallow, and R^2-values are relatively small with 0.20–0.36, indicating only low agreement of the results and the linear line fit. The reduction of the friction coefficient with higher contact pressures could also be due to a reduced scatter of the results for higher contact pressure. A one-way ANOVA of the test results with face grain showed a statistically significant difference in friction coefficients between different contact pressures (F(3, 80) = [13.85] and p < 0.0001). The results of the tests with 0.5, 1.0 and 2.5 N/mm^2 are all significantly different from those with 6.0 N/mm^2. The ANOVA for the tests with end grain show similar results (F(4, 52) = [4.80] and p = 0.0023). Here, the tests with 0.5 and 1.0 N/mm^2 significantly differ from the results with 6.0 N/mm^2.

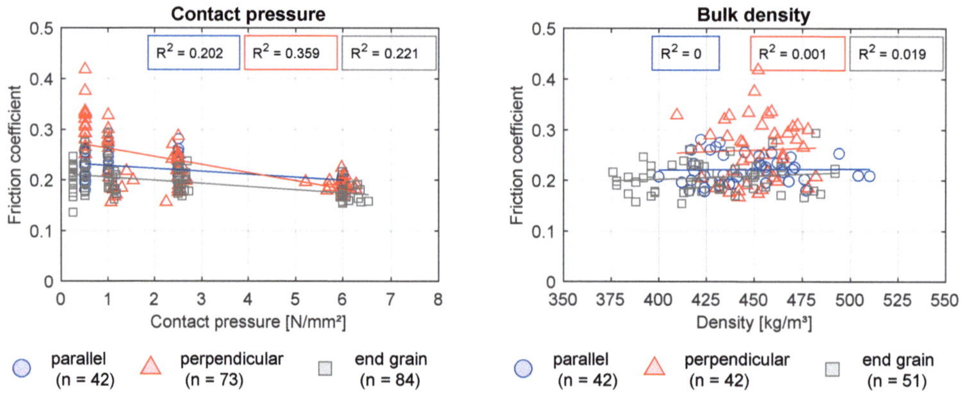

(a) (b)

Figure 2.22: Influence of (a) contact pressure and (b) density on coefficient of friction.

Influence of density

The influence of the density of the softwood was also investigated. For most tests, the density of the softwood specimens was recorded. Figure 2.22b shows the friction coefficients plotted against the density. The range of tested densities was more extensive for the tests with face grain parallel and end grain than for tests with face grain perpendicular. Still, for all three grain directions, no influence of the softwood density on the coefficient of friction can be detected.

Influence of test setup

Two different test setups were used, with the main difference being the load application of the contact pressure. Setup 1 (Figure 2.17) used a spindle to apply the normal force. The applied force changed during the test and was not constant over time. The load was constantly measured, and the corresponding load pairing determined the friction coefficient. This test setup can be assigned to category (d), like the tests with pre-stressed rods that also encountered the load-loss during tests.

Setup 2 (Figure 2.18) used a horizontal hydraulic cylinder to apply the normal force. The applied force was constant over time as the cylinder was load-controlled. Again, the load was constantly measured, and the friction coefficient was determined with the corresponding load pairing. This setup can be assigned to category (e) with two hydraulic cylinders.

Figure 2.23 shows the mean values for both setups with error bars, for all tests with all grain directions. Test setup 2 produces slightly higher results than test setup 1. However, the scatter for setup 2 is more prominent (which also might be due to almost three times the number of tests). A one-way ANOVA was performed for tests with face grain and revealed that there is a statistically significant difference in friction coefficients depending on the test setup ($F(1, 113) = [17.72]$ $p < 0.0001$). An ANOVA for tests with end grain revealed that there is also a statistically significant difference depending on the test setup ($F(1, 82) = [7.20]$ $p = 0.0088$).

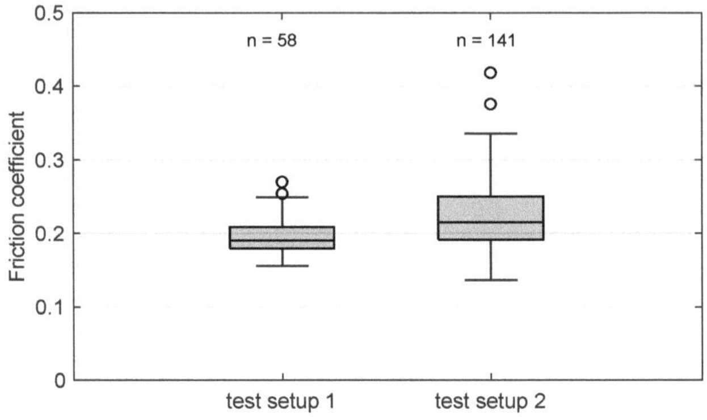

Figure 2.23: Influence of test setup on coefficient of friction (all grain directions).

2.6 Conclusion

The extensive literature research revealed insights into friction concerning timber structures. A database with a total of 3648 entries was created. This database consists of 1678 values for friction of wood on wood (static and kinetic) and 1970 values for friction of steel on wood (static and kinetic). With the collected data, various influencing parameters were investigated.

The main conclusion is that the coefficient of friction shows extreme scattering. This scattering made all attempts to compare and evaluate the collected data very difficult. Also, there is no consensus about uniform test conditions and input parameters, making comparing results from different resources difficult. Therefore, an additional parameter study was performed with 200 tests, investigating four parameters: 1) grain direction, 2) sliding speed, 3) contact pressure and 4) density.

Parameters with no influence on the friction coefficient:

From the collected data and the parameter study, it can be concluded that the coefficient of friction is independent of the grain direction of the wood for face grain. It should be differentiated between face grain and end grain, as there is a slight difference.
Again, the literature review and the parameter study showed that the coefficient of friction is independent of the sliding speed of the test.
The coefficient of friction is (mainly) independent of the contact pressure. However, at the extreme ends of the pressure range, the friction coefficients or tests to determine the friction coefficients are influenced by the contact pressure. With high contact pressure,

high stresses perpendicular to the grain of the softwood specimens occur, leading to deformation or failure of the softwood. Thus, the recorded load is not the actual maximum load of the friction system.
From the literature review and parameter study, it can be concluded that the coefficient of friction is independent of the density of the softwood.

Parameters with influence on the friction coefficient:

The moisture content of the wood has the greatest influence on the coefficient of friction. A moisture content > 20% leads to a significant increase in the friction coefficient. However, as moisture contents > 20% should be avoided in timber structures, this influence is herein of no further concern. For future investigations, it should be considered to precisely investigate the influence of an steadily increasing moisture content on the friction coefficient.
The reviewed literature shows a slight difference in the coefficients of friction determined depending on the test setup. The statistical evaluation of the parameter study shows a significant difference between the two test setups analysed. As the respective test setup specifies certain parameters or their range, the subsequent application of the analysed test specimens should be considered.
Finally, it can be concluded from the collected data that the surface condition significantly influences the friction coefficient, which was expected. This influence has excellent potential for the improvement of timber connections with dowel-type fasteners, where the rope effect increases the load-carrying capacity. The coefficient of friction can be significantly increased with appropriate surface modification. Various surface modifications are presented in the following chapter.

Recommendations

An overview of the coefficients of friction found in the literature is given in Tables 2.1–2.3. The given mean values can be used for analytical and numerical investigations. Especially the mean values for tests with either softwood or steel are based on many test results. Table 2.1 lists static and kinetic friction coefficients for softwood on softwood and hardwood on hardwood. Table 2.2 lists friction coefficients for softwood and hardwood on steel or aluminium. Table 2.3 lists friction coefficients for engineered wood products.

The given friction coefficients in the tables are for a moisture content of $u < 20\%$. For the sake of clarity (and as the evaluation of data suggests), it was not distinguished between different grain directions of the wood, and different surfaces of the steel or aluminium in Tables 2.1–2.3. The number of values taken from the literature is also given. This number does not necessarily correspond to the tests carried out. For example, in [12] a total of 9000 tests are mentioned, but only ten values can be taken from the two-page article.

3 Surface modification

3.1 Introduction

To increase the friction in the shear plane of mechanical timber connections, various types of surface modifications were investigated. The focal points of this investigation involved the connection between two elements, such as a connector and a timber member. By increasing the friction and thus the rope effect, the load-carrying capacity of this connection can be increased. Most of the surface modifications were carried out during the research project "Connectors made of densified veneer wood with increased friction in the shear plane" [3]. Therefore, densified veneer wood (DVW) was used as substrate in these tests. The properties of DVW and its use for system connectors are presented in detail in [3]. The modification methods are, of course, also applicable to other materials such as steel or aluminium. However, the results are only partially transferable.

The classification of these surface modifications adhered to the German standard DIN 8580 [107] *Manufacturing processes – Terms and definitions, division.* The nomenclature for manufacturing processes and surface modifications aligns with DIN 8580, with the presented surface modifications falling into three of the six main groups: Group 2 *Forming*, Group 3 *Parting*, and Group 5 *Coating*. Figure 3.1 provides an overview of the manufacturing processes.

The primary emphasis was on investigating milled surface modifications for system connectors. The rationale behind this approach was seamlessly integrating surface modifications into the regular manufacturing processes. Hence, various milling processes were investigated. For sheet metal components like angle brackets, the surface modification was also intended to be seamlessly incorporated into the manufacturing process, leading to the exploration of punched surfaces. It was aimed to advance the practical application of these surface modifications within the broader scope of connections in timber structures by integrating surface modifications into the manufacturing process.

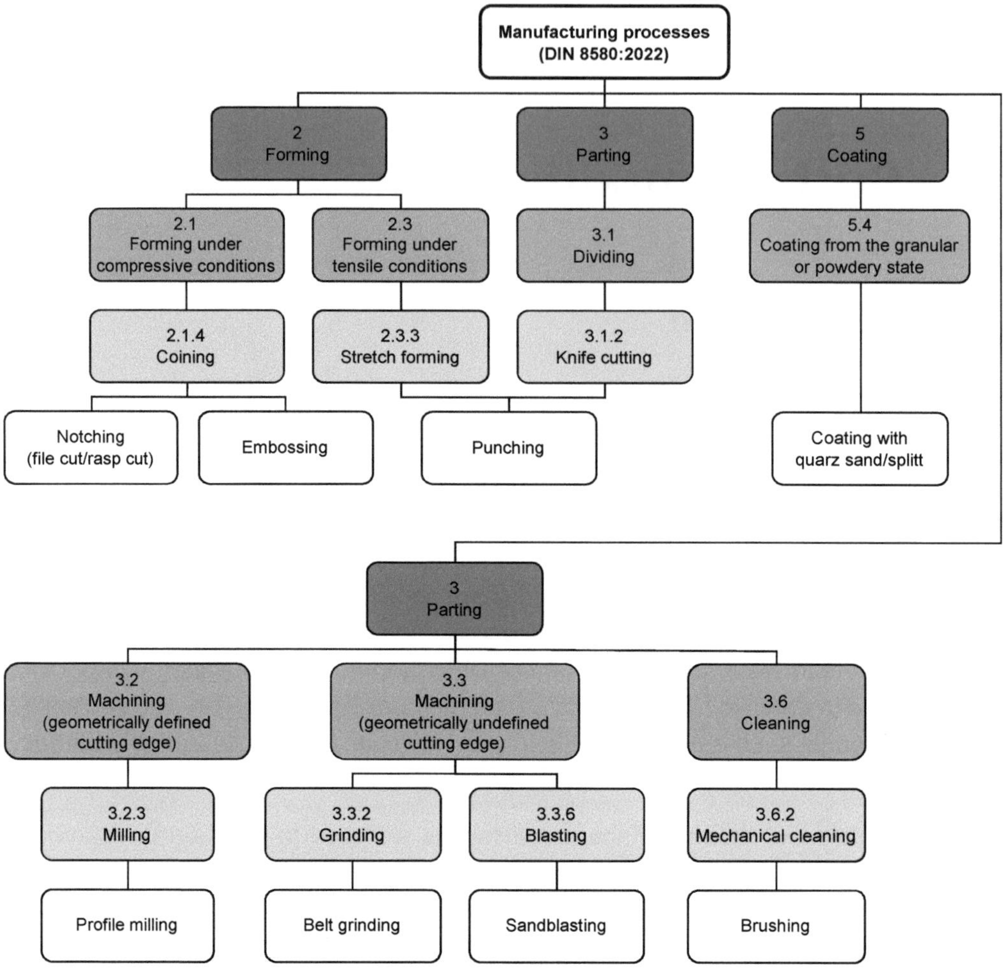

Figure 3.1: Classification of manufacturing processes according to DIN 8580.

3.2 Notching

Notching is categorized within Group 2.1 *Forming under compressive conditions*. This specific surface modification technique utilizes a punching tool to cold-form the material, see Figure 3.2a. This process results in the generation of sharp, generally parallel teeth, commonly referred to as a “file cut”, or distinct, individual teeth known as a “rasp cut”. The experimentation involved the assessment of metal plates featuring both file cut (Figure 3.2b) and rasp cut (Figure 3.2c) configurations.

It is essential to note that this surface modification is exclusive to materials that can undergo cold-forming, such as steel or aluminium. The process involves the intricate manufacturing of teeth patterns and is currently limited in its applicability to narrow workpieces. The complexity of the manufacturing process, coupled with its current feasibility solely for specific materials and narrow workpieces, underscores the nuanced nature of this surface modification technique.

File cut

A file is a tool to remove small amounts of material from a workpiece. It is commonly used in woodworking and metalworking. Most are hand tools made from a case-hardened steel rod of rectangular cross-section, with one or more surfaces cut with sharp, generally parallel teeth. The general manufacturing process of the file cut with a punching tool can be seen in Figure 3.2a. For a double-cut, the steel rod is rotated after the first set of parallel teeth, forming diamond-shaped cutting surfaces. Figure 3.2b shows the final surface of the file cut tested herein.

Rasp cut

A rasp represents a specialized type of file distinguished by its distinct, individually cut teeth designed for the coarse removal of substantial amounts of material. Much like a conventional file, manufacturing a rasp involves using case-hardened steel rods characterized by a rectangular cross-section. What sets the rasp apart is the meticulous process through which each distinct tooth is formed individually. Figure 3.2c shows the final surface of the rasp cut tested herein.

(a) (b) (c)

Figure 3.2: Notching: (a) punching tool [108]; (b) file cut; (c) rasp cut.

3.3 Embossing

Embossing is also categorized within Group 2.1 *Forming under compressive conditions*. This distinctive surface modification technique involves the utilization of a punching tool for cold-forming. In this process, a chosen pattern is impressed into the surface of the workpiece, see Figure 3.3a. This work investigated an embossed negative impression of the pyramid pattern, as shown in Figure 3.3b+c. Another variant of embossed surface investigated during this study was that of a conventional chequered plate, shown in Figure 3.3d.

Notably, this form of surface modification is versatile in its applicability, extending to various wood products, and steel or aluminium. Once the embossing punch is crafted (e.g. through milling), the embossing process is relatively straightforward, rendering it an accessible method for generating textured surfaces on diverse materials. The outcome of this technique, exemplified by the embossed negative impressions, signifies the potential for introducing patterned surfaces to enhance the functional aspects of the modified workpieces.

Negative pyramid pattern

A pyramid pattern was milled into a steel block, see Figure 3.3b. The steel block was then pressed into the surface of the test specimen. DVW plates were modified with steel blocks featuring a pyramid pattern. The pyramid tips achieved a penetration depth of approximately 1 mm into the DVW surface, resulting in an embossed negative impression of the pyramid pattern, as shown in Figure 3.3c.

Chequered plate

The used chequered plate was initially used as sheet metal for walkways (Figure 3.3d). The embossed structure was to prevent the user from slipping. Therefore, the size of the embossed pattern is scaled for work boots and might be rather large. The manufacturing process is simple, confirmed by the large availability of chequered plates in different shapes and sizes.

(a) (b) (c) (d)

Figure 3.3: Embossing: (a) embossing punch [108]; (b) used embossing punch with pyramid pattern; (c) final surface with negative imprint; (d) chequered plate.

3.4 Punching

Punching combines two manufacturing processes, i.e. Group 2 *Forming* and Group 3 *Parting*. Further classification is *Forming under tensile conditions* and *Dividing*, already describing the manufacturing process. A punching tool simultaneously forms the workpiece and removes material. Therefore, the process of punching is only applicable to (thin) metal sheets. Punched metal sheets can be produced in large quantities once the punching tool with its needed characteristics is produced. Various geometries were investigated, such as metal sheets with a round and flat collar (Figure 3.4b) or metal sheets with a sharp and protruding collar (Figure 3.4a+c).

Perforation 1

The first investigated perforation was punched in 0.25 mm thin stainless steel sheets. Here, no material was removed during the punching process. The punching tool punched through the thin metal sheet and bent the material around the punching tool upward. The metal was not adequately cut by the punching tool but rather ruptured. This rupture resulted in very sharp teeth. Depending on the penetration depth, the size of the sharp teeth can be influenced. Figure 3.4a shows a specimen with perforation 1. All teeth were punched on the same side, resulting in a one-sided modified steel sheet.

Perforation 2

The second investigated perforation was punched in 0.5 mm stainless steel sheets. The punched holes were round, and in their middle, material was removed during the manufacturing process. The punching tool bent the material around the tool upward. The metal was adequately cut and material removed, leading to round punched holes with a collar each. Figure 3.4b shows a specimen with perforation 2. Again, all teeth were punched on the same side, resulting in a one-sided modified steel sheet.

Perforation 3

The third investigated perforation was punched in 0.8 mm thin metal sheets. Like before, the punched holes were round, and in their middle, the material was removed during manufacturing. The punching tool bent the material around the tool upward. The metal was adequately cut and material removed, leading to round punched holes with a collar each. The collar, however, was significantly sharper due to the thicker metal sheet. Figure 3.4c shows a specimen with perforation 3. This time, the teeth were punched alternating on both sides.

(a) (b) (c)

Figure 3.4: Punching: (a) punched metal sheet with sharp collar of perforation 1; (b) flat collar of perforation 2; (c) alternating collar of perforation 3.

3.5 Profile milling

Profile milling is a manufacturing process categorized in Group 3.2 *Machining with geometrically defined cutting edge*. By using different milling tools, different surface patterns can be achieved. Four different surface patterns manufactured with four different milling tools were investigated. The tools used included a chamfer milling cutter with a 90° angle (Figure 3.5a), a solid end mill (Figure 3.5b), a circular mill for longitudinal and transverse grooves (Figure 3.5c), and a face milling cutter with removable inserts for circular grooves (Figure 3.5d).

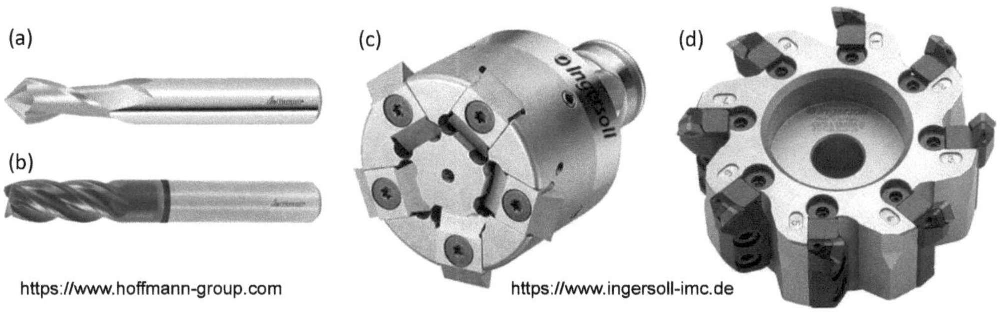

Figure 3.5: Different milling tools used for profile milling: (a) chamfer milling cutter; (b) solid end mill; (c) circular mill; (d) face milling cutter.

3.5.1 Pyramid pattern

Following [73], longitudinal grooves were milled in the top layer of the test specimens using a programmable NC router and a chamfer cutter with a taper angle of 90°. The feed rate was approx. 60 mm/min. The grooves in one test specimen were all the same depth and had the same distance from each other. The test specimen was then rotated by 90°, and parallel transverse grooves were again milled into the top layer, leaving small pyramids forming a rough surface. Test specimens with 0.5, 1.0, 1.5 and 2.0 mm deep grooves were produced. Figure 3.6 shows examples of test specimens with 0.5 mm deep grooves and with 2.0 mm deep grooves. When milling the 0.5 mm deep grooves, the problem arose that some of the pyramids chipped off during the manufacturing process, which is why the transverse grooves were then milled slightly deeper than the longitudinal grooves.

In a variation, the circular mill was used to mill the grooves. The specimens were manufactured by a project partner, and no data on the milling speed is available.

However, the same challenges arose as before. In general, this surface modification applies to most materials. However, depending on the depth of the grooves, chipping of some pyramids might occur. The main disadvantage is the complex and time-intensive manufacturing process, as every groove must be milled separately (at least in this project). For the herein manufactured specimens with a milled surface of 110x120 mm^2 the total milling time varied between 1.8 hrs for the 2 mm pyramids and up tp 3.7 hrs for the 1 mm pyramids (for one side of the two-sided specimens).

(a) (b) (c)

Figure 3.6: Milled pyramid pattern: (a) 1.5 mm; (b) 0.5 mm; (c) 2.0 mm.

3.5.2 Circular pattern

In principle, face milling can produce very flat surfaces. In practice, however, the result always shows visible trochoidal marks that follow the motion of the inserts of the cutter (see Figure 3.7a). These trochoidal marks give the characteristic finish of a face-milled surface. If cutting inserts are removed from the milling tool and the speed increased, visible and perceptible trochoidal marks remain. This concept was used for the circular milling pattern, where only two diagonal milling inserts were used on the milling tool. The remaining inserts were removed before milling. The milling tool rotated at a constant high speed and was run over the surface of the test specimens at a high feed rate. With this tool, circular grooves were milled into the top layer of the specimens, cutting themselves repeatedly. As a result, pyramid-like shapes remained towards the edge of the test specimens, while elongated grooves were in the centre of the test specimens. The milling pattern was milled into plates of DVW (Figure 3.7b) and steel (Figure 3.7c). Different cutters were used, but the results were always similar. The rotational speed and the feed rate had a significant influence on the surface quality. For the specimens with DVW, a rotational speed of 450 rpm and a feed rate of 1000 mm/min were used. The specimens with steel were manufactured with a rotational speed of 500 rpm and significantly higher feed rate of 3000 mm/min. The depth of the grooves was approx. 0.5 mm. The manufacturing process was straightforward, as the circular milling produced grooves in both directions. Also, the manufacturing time was short when compared to the pyramid pattern.

(a) (b) (c)

Figure 3.7: Circular pattern: (a) trochoidal marks; (b) circular pattern in DVW and (c) in steel.

3.5.3 Horizontal grooves

Parallel grooves were milled into the surface with the solid end mill, like the pyramid pattern. The end mill was inclined by 5°, leaving a sharp horizontal edge. The distance between the grooves was 10 mm, significantly greater than for the pyramid pattern. The manufacturing process was much faster due to the greater distance between the grooves. Figure 3.8a shows the final surface on a steel plate.

3.5.4 Scale pattern

Additional grooves perpendicular to the parallel grooves were milled into the surface using the same solid end mill as before and the same tilting angle of 5°. Like the pyramid pattern, a protruding pattern with sharp edges remained. This pattern resembled a scale pattern, as seen in Figure 3.8b. The manufacturing process increases in time due to the additional grooves compared to the horizontal grooves. However, the manufacturing time is still shorter than for the pyramid pattern due to the greater distance of the grooves.

(a) (b)

Figure 3.8: Horizontal grooves in steel (a) and scale pattern in DVW (b).

3.6 Belt grinding

Belt grinding is categorized in Group 3.3 *Machining with geometrically undefined cutting edge*. The used belt grinder was equipped with P40 grinding paper. The DVW specimens were sanded on both sides over their entire surface, similar to Figure 3.9a. The sanding was done parallel to the later load direction during the friction tests. The sanding resulted in a noticeable structuring. The manufacturing time is very short for a surface roughened to the touch.

3.7 Sandblasting

Also categorized in Group 3.3 is the process of sandblasting. Each test specimen was manually sandblasted for about two minutes per side in a small sandblasting chamber. The abrasive was normal corundum F080 with a grain size of 150–212 µm. For the specimens made of DVW, sandblasting resulted in slightly different surfaces on each side and each test specimen. During sandblasting, it was observed that the blasting material removed the early wood of the veneers, and the latewood remained. This removal resulted in structuring along the grain direction of the face veneers, thus parallel to the later load direction.

Steel and aluminium can also be sandblasted, but a homogeneous surface results due to the material's structure. This surface can be seen in Figure 3.9b, which shows a steel plate where the left side was blasted with the corundum mentioned above. The shown blasted steel plate was used for tests with coated surfaces. The manufacturing time of the sandblasted surfaces is very short.

3.8 Brushing

Brushing is categorized in Group 3.3 *Cleaning*. A cup brush with knotted steel wire was used to brush both sides of the smooth surface of the test specimens. A clear structuring of the surface can be seen in Figure 3.9c. Like the milled circular pattern, the brushing was performed with an NC milling machine. Due to the high feed rate, the manufacturing time of a brushed surface is very short.

(a) (b) (c)

Figure 3.9: Parting: (a) belt grinding [109]; (b) sandblasted (left side) steel plate; (c) brushed surface.

3.9 Coating

The final investigated group was Group 5 *Coating*. Two different types of adhesive were used for the coating, and a total of three different adhesive products. On the one hand, two flowable two-component adhesives were used, which cured at room temperature. On the other hand, an epoxy resin adhesive tape was used, which cured at temperatures of 130–170°C. The test specimens were coated with quartz sand with a grain size of 0.5–1.0 mm and 0–2 mm and with grit with a grain size of 2–4 mm (see Figure 3.10). In addition, a test series was coated with skateboard grip tape. Grip tape is an adhesive tape coated with sand grains for the top of skateboards to have a firm footing while riding and better control over the skateboard.

3.9.1 Two-component adhesive (2K-SE)

For the test specimens with DVW as the substrate, the two-component universal adhesive 2K SE-Polymer 690.00 by Jowat Adhesives [74] was used, a hybrid system of epoxy resins and silane-terminated polymers (Figure 3.10a). The surface of the test specimens was sandblasted before the adhesive was applied, analogous to the sandblasted test specimens. The adhesive was applied from a two-component cartridge with a static mixer, which ensured homogeneous mixing of the two components. An application gauge was used to ensure that an adhesive layer of constant thickness was applied to each side. For the coating with quartz sand 0–2 mm, a 0.5 mm thick adhesive layer was chosen; for the coating with grit 2–4 mm, a 1.0 mm thick adhesive layer. The test specimens were then pressed manually into the respective aggregate (Figure 3.10b). According to the manufacturer's instructions, curing occurred at room temperature for one week. Applying the adhesive on both sides was more difficult than expected because the adhesive was very liquid. Therefore, the adhesive was allowed to dry for about 15 minutes before the test specimens were pressed into the aggregate.

For the test bodies with steel as the substrate, the two-component adhesive Sikadur-370 by Sika [75] was used. The steel plates were sandblasted and degreased before the adhesive was applied. The two components of the adhesive were not mixed using a cartridge with a static mixer but were combined in advance using a centrifugal mixer. The thickness of the adhesive layer was specified on both sides of the test bodies with adhesive tapes (Figure 3.10c). Only a coating with quartz sand 0.5–1.0 mm and an adhesive layer thickness of 0.5 mm were investigated. Here, as well, the test specimens were pressed manually into the aggregate (Figure 3.10d). The grading curve of the aggregate showed a very low proportion of fine grains, with about 90% of the grain size between 0.5 and 0.8 mm. Curing was carried out at room temperature according to the manufacturer's instructions for eight days.

The manufacturing process is very complex. Much preliminary work must be done to realise a clean adhesive bond: sandblasting, degreasing, and curing. This preliminary work is all labour-intensive and time-consuming.

3.9.2 Epoxy adhesive tape (EpoxyTape)

As an alternative to the flowable or higher viscosity liquid adhesives, adhesive tapes were investigated for easier handling and processability. These were epoxy resin adhesive tapes from Lohmann Tapes: DuploTEC 10490 SBF-Epoxy Tape [76] with an adhesive layer thickness of 0.1 mm, and DuploTEC 10650 SBF-Epoxy Tape [77] with an adhesive layer thickness of 1.0 mm were used (Figure 3.10e). Both tapes were applied at room temperature and then cured in the oven at a temperature between 130–170°C.

Only the 0–2 mm quartz sand was chosen as the aggregate for both adhesive tape versions (Figure 3.10f).

Furthermore, the quartz sand was pressed on with a constant contact pressure of 2 N/mm^2 for two minutes. The clear advantage of the epoxy resin adhesive tapes is the defined layer thickness. The manufacturing time could be sped up using the tape instead of the liquid adhesive. The most significant disadvantage, however, is the thermal curing.

Figure 3.10: Differently coated specimens: (a) Jowat 2K-SE 690; (b) coated with grit; (c) Sika Sikadur-370; (d) coating with sand 0.5–1.0 mm; (e) Lohmann DuploTEC 10650; (f) coated with sand 0–2 mm.

3.9.3 Grip tape

A commercial grip tape, which is used for the top of skateboards for better adhesion, was used. The grain of the grip tape was much finer than the silica sand and resembled sandpaper (Figure 3.11a+b). The application of the grip tape was much easier than that of the two epoxy adhesives. Manufacture-wise, this was the most convenient coating. The tape was easy to handle, and the sand grains were already applied to the tape.

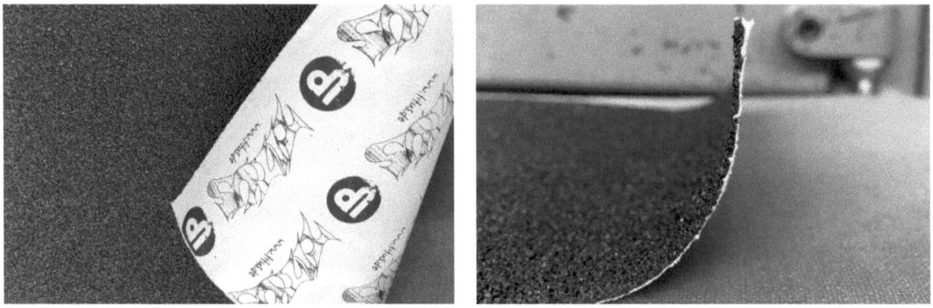

(a) (b)

Figure 3.11: Grip tape: (a) front and back view; (b) side view.

3.10 Conclusion

Eight different modification processes were investigated, resulting in 17 different surfaces. The surface modifications ranged from simple methods, such as belt grinding and sandblasting, to more complex methods, such as profile milling and coating with sand.

Notching is a process where specific cuts or notches are made in a material. The here used notched surfaces were commercially available files and rasps. While the surface modification of metal plates with notching is feasible, the width limitation suggests potential customisation or scalability constraints.

Embossing, on the other hand, is described as a relatively straightforward process. With an embossing punch, the negative imprint of the tool is embossed into the surface. Once the initial effort is invested in creating the embossing punch, the possibilities become extensive, and customisation in terms of size and design is achievable. This flexibility is a notable advantage, making embossing an attractive option for modifying surfaces.

Punching is considered a viable option, particularly for products made of sheet metal, such as angle brackets and joist hangers. During the punching process, the material is removed by a punching tool. Depending on the punching tool, a collar or similar may remain at the punched hole. Angle brackets and the like are already punched out of sheet metal during manufacturing, implying that the surface modification integrates well into the manufacturing process. The surface modification with punching is feasible even in larger sizes. The efficiency and potential for scalability make punching a very promising surface modification.

Commercially available system connectors are usually milled from a single blank. Milling is time-consuming, especially with the examined tools, which mill each groove individually. However, milling becomes promising with the appropriate tools. Additionally, it can be incorporated into the initial manufacturing process of system connectors. Horizontal grooves only in one direction are proposed as an alternative, particularly when load transfer occurs mainly in one main direction. A circular pattern is deemed easy to produce, but challenges such as strain on the machine and a lack of optimal cutting tool inserts are noted. The suggestion of a scale pattern as an interim solution implies a pragmatic approach to balancing efficiency and effectiveness during development.

Belt grinding is a simple and effective method, indicating its practicality in quickly achieving desired results. Similarly, sandblasting and brushing are also described as simple and effective techniques. If the brushing process is further improved (e.g. type of brush, feeding rate, inclination angle), it could pose an alternative to circular milling, combining the speedy manufacturing process with a promising surface structure and reducing the strain on the milling machine and cutting inserts.

Coating is acknowledged as a time-consuming process with substantial preparatory work. However, its efficiency, when executed correctly, is emphasised. The careful selection of grain size highlights the precision required in the coating process for optimal results. This description underscores the importance of meticulous planning and execution to efficiently achieve the desired surface modification.

In summary, insights into the various surface modification techniques and their feasibility, efficiency, and potential considerations for optimisation in a manufacturing context were provided.

4 Friction tests

4.1 Introduction

Friction tests were carried out to quantify the different surface finishes. Tests were carried out with DVW on softwood and with steel on softwood to determine the coefficient of friction between modified surfaces and softwood. In total, 26 different surfaces were investigated, and 844 tests were evaluated. For all tests, the static coefficient of friction and the kinetic coefficient of friction were determined. Statistical analysis was used to determine whether friction coefficients for various surfaces differ. Additionally, 5%-quantile values were determined. The results of the tests are given in Tables 4.2–4.12 in this chapter and Appendix A.1. However, the data in these tables should not be applied to situations significantly different from those used to obtain them [27].

4.2 Test setup and execution

The experimental setup for the tests to determine the friction coefficients is explained in detail in Chapter 2 Section 2.5.1. Two different test setups were used, but no significant difference was distinguished. Therefore, no further distinction is made in the following. In test setup 1, the normal force F_n, perpendicular to the friction surface, was applied with a threaded rod and a spindle. In test setup 2, F_n was applied with a hydraulic cylinder. The frictional force F_f parallel to the friction surface was applied to the test specimens displacement-controlled, up to a displacement of 15 mm (of the machine head), with a universal testing machine.

During the tests, softwood face grain and end grain were distinguished. Furthermore, the grain direction was distinguished between parallel or perpendicular to the loading direction. This resulted in a friction surface of 100×100 mm^2 for the tests with face grain and a surface of 50×100 mm^2 for the tests with end grain. The size of the friction surface was chosen to resemble the size of a connector plate and, therefore, the size in a later connection. The contact pressure in the friction surface was the same for all tests of one series. That was the case for the tests with setup 1, at least at the start of the test, as the spindle could not constantly hold the contact pressure. For the tests with setup 2, the contact pressure was constant throughout the tests, as the hydraulic cylinder was load-controlled.

Based on the results of the parameter study in Chapter 2 Section 2.5, a test speed of 5 mm/min and a contact pressure of 2.5 N/mm^2 were chosen for the tests. For both test setups, the machine load, the machine displacement, and the horizontal load were measured continuously with a measuring rate of 100 Hz.

The test specimens for most tests were thick enough to push on them directly. Clamping jaws were used to pull on the steel sheets for the tests with thin steel sheets. For these tests, a slippage of approximately 3 mm was observed between the machine head and the specimens (presumably because of some movement of the clamping jaws). The slippage was not subtracted in the results and should be kept in mind when evaluating the load-displacement diagrams. For the tests with specimens that were pushed on, it was not explicitly checked for slippage.

The static coefficient of friction was calculated according to Equation 4.1 as the ratio of frictional force F_f to the normal force F_n. The frictional force F_f parallel to the contact surface was divided equally between the two friction surfaces. Three typical behaviours of surfaces are shown in Chapter 2 Figure 2.2, and $F_{f,static}$ was chosen accordingly. In cases where stick-slip occurred, the load $F_{f,static}$ was chosen as the maximum force at the very first peak, according to American standard ASTM G115.

The kinetic coefficient of friction was calculated with the force $F_{f,kinetic}$ corresponding to the horizontal part of the load-displacement curve. As that part was not always explicit (slope of the curve still increasing/decreasing or any other form of non-linear behaviour), the kinetic coefficient of friction was evaluated as the mean value between a displacement of 10 to 15 mm. The range of 10 to 15 mm was chosen as this was the range where most curves ran horizontally. With so many tests and sometimes quite different curves in one series, it was impossible to evaluate the friction coefficients automatically with a MATLAB®-script. Therefore, some friction coefficients had to be evaluated manually, contrary to the approaches mentioned.

$$\mu = \frac{F_f}{2 \cdot F_n} \tag{4.1}$$

Table 4.1 gives an overview of all investigated surfaces and all performed tests. The surfaces are sorted and numbered according to the manufacturing processes presented in Chapter 3. In total, 844 tests were carried out with 26 different surfaces. The test matrix was extended on-the-fly. While only the face grain perpendicular and end grain were examined for the first tests, later tests with face grain parallel were added. Additionally, the number of tests for each series was decided based on the data available up to that point (which not only accounted for the friction coefficient but also, e.g. for the manufacturability of the surface).

Table 4.1: Overview of the experimental programme.

No.	Surface	No. of tests n			Test setup
		Face grain ∥	Face grain ⊥	End grain	
0	Untreated				
0.1	DVW	42	73	84	1 + 2
0.2	Aluminium	12	12	12	1 + 2
1	Notching				
1.1	File cut	8	8	8	2
1.2	Rasp cut	11	11	8	2
2	Embossing				
2.1	Inverse pyramid pattern	10	30	21	1
2.2	Chequered plate	13	13	14	1 + 2
3	Punching				
3.1	Perforation 1	-	3	-	1
3.2	Perforation 2	5	5	5	2
3.3	Perforation 3	10	10	10	2
4	Milling				
4.1	Pyramid pattern				
4.1.1	0.5 mm	10	40	30	1
4.1.2	1.0 mm	20	15	10	1
4.1.3	1.5 mm	19	20	18	1
4.1.4	2.0 mm	-	12	-	1
4.2	Circular pattern				
4.2.1	DVW	20	14	13	1
4.2.2	Steel	12	12	24	2

Continued on the next page

Table 4.1 – continued from previous page

No.	Surface	No. of tests n			Test setup
		Face grain ∥	Face grain ⊥	End grain	
4.3	Horizontal pattern	5	5	5	2
4.4	Scale pattern	10	10	5	2
5	Belt grinding	-	6	6	1
6	Sandblasting	-	6	6	1
7	Brushing	-	5	5	1
8	Coating				
8.1	2K-SE				
8.1.1	Sand 0.5–1.0 mm	4	3	3	2
8.1.2	Sand 0–2 mm	-	3	3	1
8.1.2	Grit 2–4 mm	-	3	3	1
8.2	EpoxyTape				
8.2.1	Tape 0.1 mm	-	3	3	1
8.2.2	Tape 1.0 mm	-	5	3	1
8.3	Griptape	-	3	4	1

For most tests, the friction force F_n was applied by pushing on top of the specimens, thus compression. This was done because it was the easiest way to do it in terms of the size of the specimens, complexity of the setup, and ease of handling the specimens. For the tests with the punched steel plates, the load was applied by pulling on the specimens, thus creating tension. This was done because the thin steel plates could only be pulled. The setup, however, was more complex: the size of the specimens had to be almost twice as long, and each timber part had to be drilled for the hold-downs. Differences that have arisen are briefly discussed below. Figure 4.1a+b shows the interlocking of the modified surface with the timber side members during the tests with specimens that were pushed throughout the friction test. The twisting of the side members and consecutive interlocking of the surfaces could have led to higher friction coefficients. Exemplarily for a test with end grain, Figure 4.1a shows the twisting of the top of the side members. Consequently, tensile failure perpendicular to the grain occurs, and cracks become visible. This behaviour is not as pronounced in tests with face grain parallel, as can be seen in Figure 4.1b. But still, interlocking occurs, as well as tensile failure perpendicular to the grain of the side members.

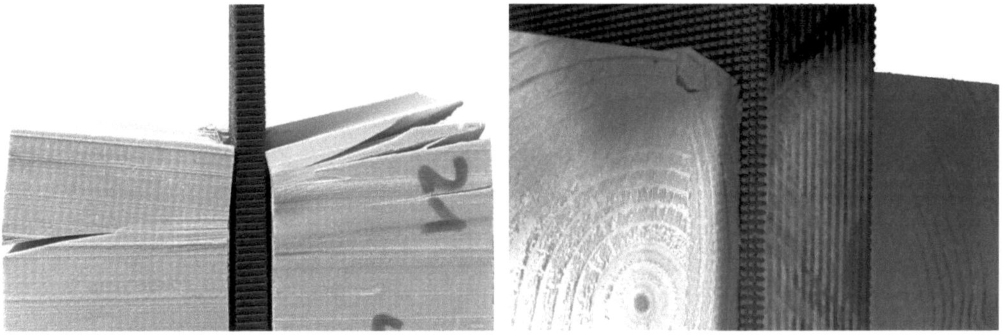

(a) (b)

Figure 4.1: Interlocking of the modified surfaces with the timber members, exemplarily for tests with end grain (a) and face grain parallel (b).

This could not be observed to such an extent in the tests with the steel plates under tension. Theoretically, interlocking can also occur in tests under tension (now at the bottom of the specimens), as the setup is mirrored. Therefore, future research should further evaluate this case with comparative tests to see which setup (push or pull) produces more realistic results.

Also, the option of single-sided friction tests should be further researched. As stated before, the double-sided test setup was chosen for its simplicity. However, as shown, interlocking occurs, leading to possible higher friction coefficients. A single-sided setup might prevent the interlocking effect, potentially leading to more authentic friction coefficients. In a single test with a single-sided setup (beech LVL on softwood), no difference in the friction coefficient could be determined.

To perform single-sided friction tests, the opposite side of the modified surface must be frictionless. This can be done with, e.g. PTFE (polytetrafluoroethylene, commonly known as Teflon) or PP (polypropylene). However, internal tests have shown that the friction coefficient of PP on PP is not necessarily zero (or close to zero). Only when multi-purpose oil was added to the friction plane was a friction coefficient of almost zero determined. Alternatively, a setup with roller bearings can be considered. The necessity of a single-sided setup must be investigated in comparative tests.

4.3 Results and discussion

In this chapter, the results of the friction tests are presented. For each test, the static and the kinetic friction coefficient were evaluated. Figure 4.2a shows a typical COF-displacement curve. Highlighted in red are the static friction coefficient at the first peak and the kinetic friction coefficient as the average between 10–15 mm of displacement. Figure 4.2b shows the corresponding force-displacement diagram of F_f and F_n.

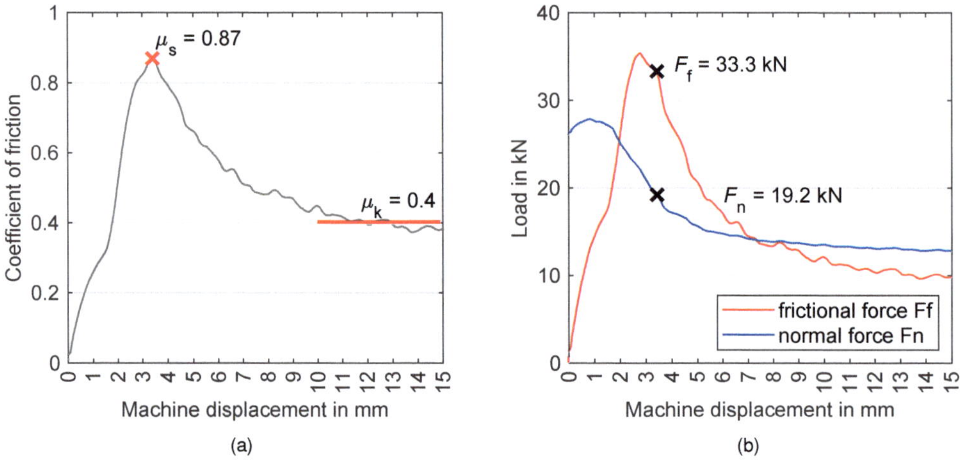

Figure 4.2: Typical curve for friction coefficient over machine displacement (a) and load-displacement curves for frictional force F_f (red) and normal force F_n (blue) of the same test (b).

4.3.1 Untreated

Densified veneer wood (DVW)

The tests with untreated DVW and softwood showed a pronounced stick-slip behaviour, especially with face grain and low sliding speed (v = 1 mm/min, see Figure 4.3a). Stick-slip behaviour no longer occurred in the tests at high speed (v = 10 mm/min, Figure 4.3b), which was to be expected (see explanation in Chapter 2, Section 2.1). The load-displacement curves followed diagram 1 (Figure 2.2a) for the tests with high sliding speeds and diagram 2 (Figure 2.2b) for the tests with low sliding speeds. The tests with smooth DVW and face grain parallel resulted in a mean value of μ_s = 0.22 and with face grain perpendicular of μ_s = 0.23. Additionally, the tests with end grain only reached μ_s = 0.20. Thus, the mean values were even lower than the value of 0.25 that is used for the rope effect according to Eurocode 5.

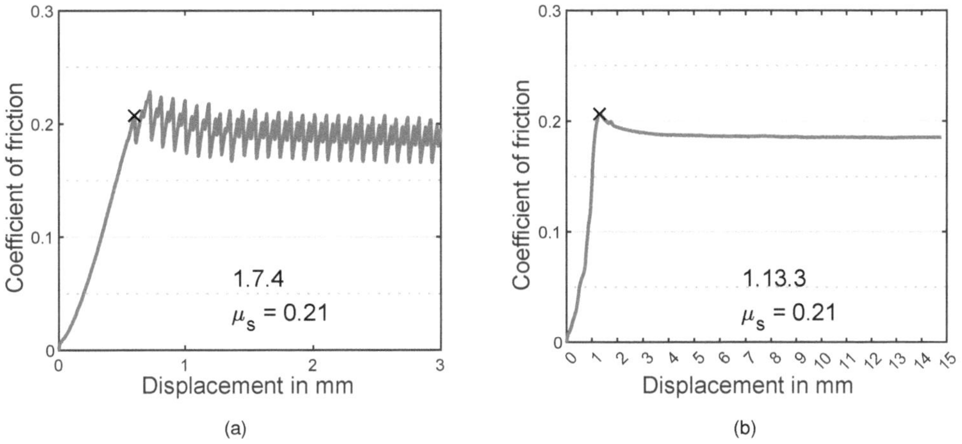

Figure 4.3: Tests with untreated DVW and low sliding speed (a) and high sliding speed (b).

In none of the tests with a contact pressure of σ_N = 1–2.5 N/mm^2 the DVW or the softwood were damaged. However, the tests with a contact pressure of σ_N = 6 N/mm^2 failed due to compression perpendicular to the grain. Therefore, the size of the DVW test specimens was reduced from a width of 100 mm to 25 mm, see Figure 4.4. This size reduction still resulted in a compression failure perpendicular to the grain, but the contributing fibre length could be significantly increased, and a coefficient of friction could be determined. The results for the static and kinetic friction coefficients are given in Table 4.2. As the literature review and the parameter study with the untreated DVW showed no influence of the contact pressure on the friction coefficient, it was not further distinguished between the different investigated contact pressures for the untreated DVW. The same goes for the two different test setups, used for the tests with untreated DVW.

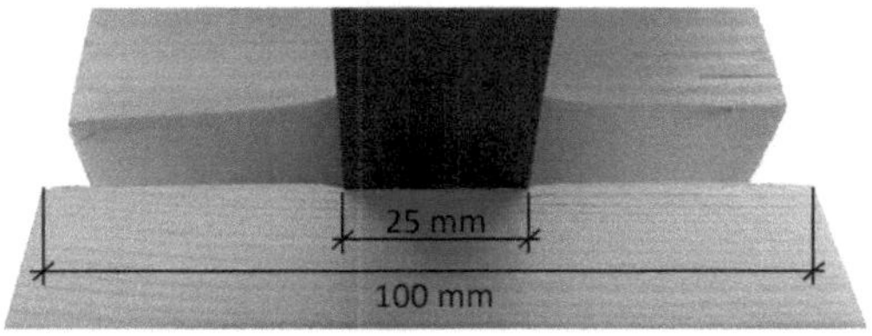

Figure 4.4: Tests with untreated DVW and contact pressure σ_N = 6 N/mm^2.

Anodised aluminium

Tests were also carried out with anodised aluminium (and softwood), as used for commercially available system connectors, to quantify the results of the other surface modifications. The tests were performed with actual connector plates; see Figure 4.5. The results averaged μ_s = 0.25 and μ_s = 0.34 for the face grain tests parallel and perpendicular, and μ_s = 0.39 for the end grain tests. Interestingly, aluminium was the only surface following diagram 3 (Figure 2.2c) with a rising slope after reaching $F_{f,static}$. Therefore, the given kinetic coefficient of friction is only a snapshot of the test results with a maximum displacement of 15 mm. All results, including the kinetic coefficients of friction, are given in Table 4.2.

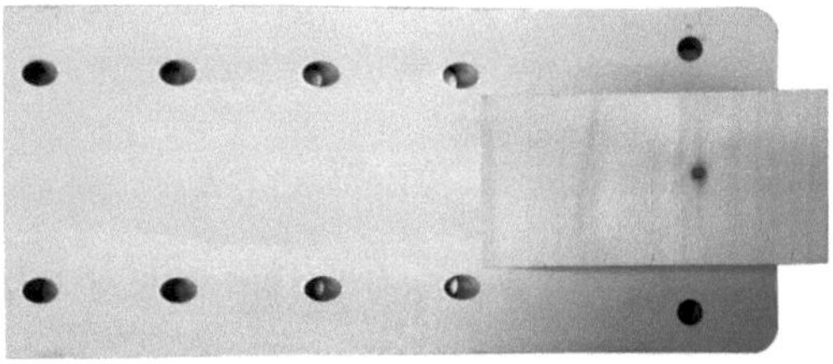

Figure 4.5: Tests with aluminium.

As seen in Table 4.2, the untreated DVW values are lower than aluminium values. That is because the surface of the used DVW was smooth and resembled a lacquered wood surface, whereas the used aluminium was anodised and already had a rougher feel to the surface.

Table 4.2: Coefficient of friction for untreated surfaces (DVW and aluminium) and softwood.

		Face grain ‖		Face grain ⊥		End grain	
		μ_s	μ_k	μ_s	μ_k	μ_s	μ_k
DVW	MEAN	0.22	0.19	0.23	0.19	0.20	0.19
	SD	0.03	0.02	0.06	0.03	0.03	0.03
	COV	12%	8%	25%	17%	15%	15%
Aluminium	MEAN	0.25	0.24	0.34	0.41	0.39	0.45
	SD	0.05	0.06	0.06	0.10	0.05	0.02
	COV	20%	23%	17%	24%	13%	4%

MEAN = mean value SD = standard deviation COV = coefficient of variation

4.3.2 Notched

File cut

High friction coefficients were achieved with notched steel plates with a file cut and softwood. A mean value of μ_s = 1.84 was evaluated for the tests with face grain parallel. Mean values of μ_s = 1.06 and 1.11 were evaluated for the tests with face grain perpendicular and end grain. The load-displacement curves mainly followed diagram 1 (Figure 2.2a). The test specimens showed significant damage afterwards. Especially for the tests with end grain, the file cut interlocked with the softwood, leading to tensile failure perpendicular to the grain of the softwood specimens. This leads to the assumption that an interlocking of the surfaces obviously works better with face grain than end grain, due to the different mechanical properties of the wood according to the fibre direction. Figure 4.6 shows test specimens after the test.

(a)

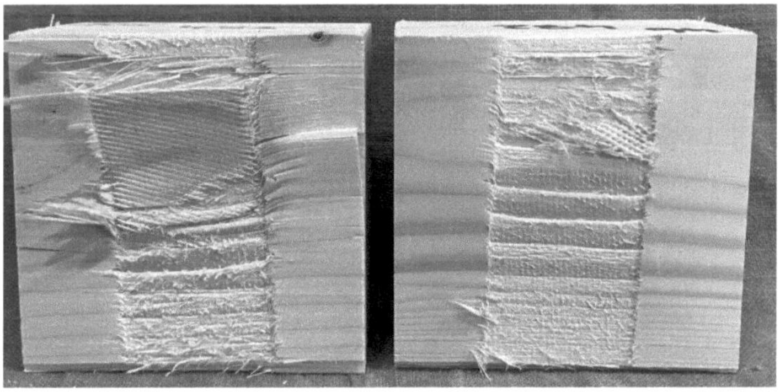

(b)

Figure 4.6: Tests with file cut: (a) wood fibres sticking to file; (b) damaged softwood specimens.

Rasp cut

By far, the highest friction coefficients were achieved with the steel plates with rasp cut and softwood with face grain parallel. A mean value of μ_s = 2.60 was determined, and for the tests with face grain perpendicular, a mean value of 1.33. However, the high friction coefficients come with high coefficients of variation. The load-displacement curves mainly followed diagram 1 (Figure 2.2a). Severe damage to the test specimens occurred during the tests. The softwood specimens parallel to the grain showed shear failure of the top layers of fibres (equal to the depth of the rasp cut), and the specimens perpendicular to the grain showed the significant interlocking effect of the rasp cut, combined with rolling shear failure of the top fibre layers. The end grain specimens showed compressive failure perpendicular to the grain. Except for wood fibres sticking to the rasp cut, no damage to the steel surface was observed. The failure behaviour, coupled with the high friction coefficients, makes it difficult to regard the results as explicit friction coefficients. Nevertheless, the potential of the notched surface modification, especially the rasp cut, is evident. All results, including the kinetic coefficients of friction, are given in Table 4.3.

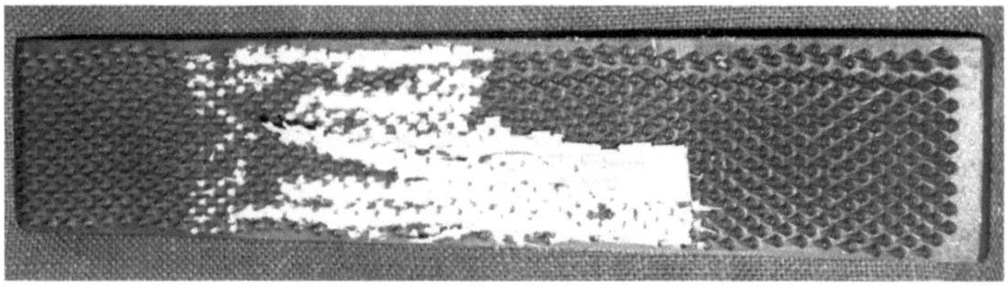

(a)

(b)

Figure 4.7: Tests with rasp cut: (a) wood fibres sticking to rasp; (b) damaged softwood specimens.

Table 4.3: Coefficient of friction for notched steel and softwood.

		Face grain ‖		Face grain ⊥		End grain	
		μ_s	μ_k	μ_s	μ_k	μ_s	μ_k
File cut	MEAN	1.83	0.60	1.06	0.61	1.11	0.86
	SD	0.27	0.06	0.12	0.06	0.23	0.15
	COV	15%	10%	11%	9%	21%	18%
Rasp cut	MEAN	2.60	0.69	1.33	0.74	1.47	1.37
	SD	0.61	0.14	0.40	0.05	0.12	0.22
	COV	23%	20%	30%	7%	8%	16%

4.3.3 Embossed

Negative pyramid pattern

For the embossed surface of the DVW with inverse pyramid pattern, mean values of $\mu_s = 0.79$ for face grain perpendicular and $\mu_s = 0.71$ for end grain could be determined. The friction coefficient for face grain parallel was slightly lower with $\mu_s = 0.67$. All three grain directions show similar COV values, ranging from 8–17%. The load-displacement curves mainly followed diagram 1 (Figure 2.2a). Interestingly, there was severe damage to the surface of the softwood specimens with fibres sticking to the DVW, see Figure 4.8a, although the modified surface was relatively flat with no protruding features. Figure 4.8b shows the influence of already tiny knots on the surface and the damaging effect due to the locally very high density. The embossed pattern is easy to manufacture and shows excellent potential. The results indicate that surfaces do not necessarily have to have protruding features to achieve adequate friction coefficients.

(a) (b)

Figure 4.8: Embossed DVW surface after the tests: (a) damaged surface of DVW; (b) influence of knots on surface.

Chequered plate

The tests with the chequered steel plate and softwood reached mean values of $\mu_s = 0.66$ for face grain parallel, and lower values for face grain perpendicular and end grain with $\mu_s = 0.59$ and 0.58. The load-displacement curves mainly followed diagram 3 (Figure 2.2c) with no significant peak, making it difficult to determine the "right" static friction coefficient. Clear imprints of the pattern can be recognised in the face grain of the softwood; see Figure 4.9. It is, therefore, astonishing that the friction coefficients are so low. However, this also shows that protruding surface features must be restricted in size to be effective for connections. All results, including the kinetic coefficients of friction, are given in Table 4.4.

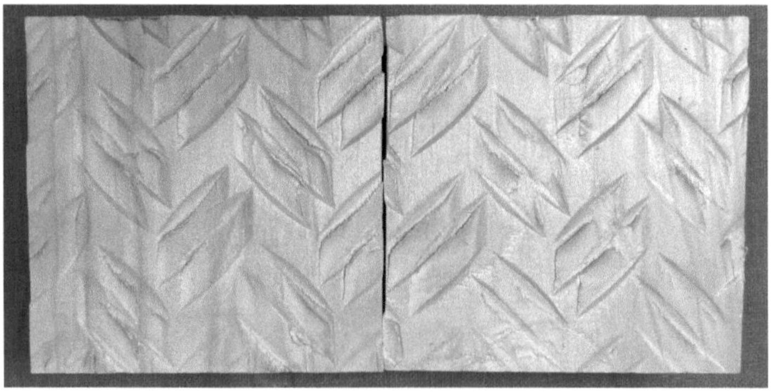

Figure 4.9: Imprint of chequered plate in softwood.

Table 4.4: Coefficient of friction for embossed surfaces and softwood.

		Face grain ‖		Face grain ⊥		End grain	
		μ_s	μ_k	μ_s	μ_k	μ_s	μ_k
Inverse pyramid pattern	MEAN	0.68	0.50	0.79	0.52	0.71	0.52
	SD	0.09	0.06	0.10	0.12	0.06	0.07
	COV	14%	11%	13%	22%	8%	13%
Chequered plate	MEAN	0.66	0.65	0.59	0.61	0.58	0.64
	SD	0.20	0.15	0.18	0.17	0.04	0.04
	COV	31%	24%	30%	28%	7%	6%

4.3.4 Punched

Perforation 1

The metal sheets with perforation 1 had the sharpest punched collar. However, this was not mirrored in the results. The only tests with softwood face grain perpendicular resulted in an average value of $\mu_s = 0.82$. The curves followed diagram 1 (Figure 2.2a). The metal sheets had a thickness of 0.25 mm. In the first tests, tensile failure of the steel sheets occurred, and no friction coefficient could be determined. Therefore, the contact pressure was reduced to approximately 1.0 N/mm^2. The size of the softwood specimens was also reduced to 50x100 mm. Figure 4.10 shows the imprint of the sharp punched collars.

Figure 4.10: Imprint of punched perforation 1 in softwood.

Perforation 2

The metal sheets with perforation 2 with its evenly spaced circular punched collars reached average values of μ_s = 0.78 for tests with face grain parallel and significantly higher values with μ_s = 0.98 for face grain perpendicular. The COV values are very low for both grain directions; however, only two tests were performed for each. The curves followed diagram 3 (Figure 2.2c), with the tendency of a decreasing slope. No damage to the surface or the softwood was observed.

Figure 4.11a shows the softwood surface after the tests with face grain parallel and Figure 4.11b with face grain perpendicular. Some impression of the collars can be seen, but mostly the perforated metal sheets slides along the softwood surface.

(a)

(b)

Figure 4.11: Softwood specimens after tests with metal sheets with punched perforation 2 and face grain parallel (a) and face grain perpendicular (b).

Perforation 3

The metal sheets with perforation 3 resembled the previous perforation with a similar circular pattern. However, the punched collars had a sharp edge from the manufacturing process. The average values for face grain parallel and perpendicular were alike with $\mu_s = 0.90$ and 0.94, respectively. The curves followed mainly diagram 1 (Figure 2.2a) with a clear peak of F_f. The metal sheets were undamaged after the tests. In addition, the punched collars did not lose any of their "sharpness", and multiple tests could be performed with the same metal sheet. The surface of the softwood with face grain parallel to the sliding direction is shown in Figure 4.12a and face grain perpendicular to the sliding direction in Figure 4.12b. The interlocking with the wood fibres can be seen clearly. All results, including the kinetic coefficients of friction, are given in Table 4.5.

(a)

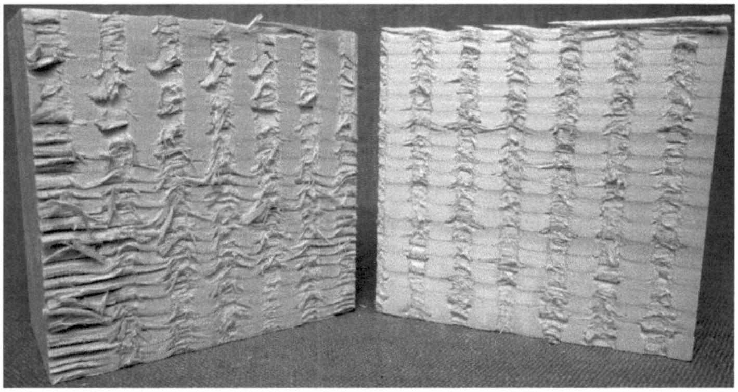

(b)

Figure 4.12: Softwood specimens after tests with metal sheets with punched perforation 3 and face grain parallel (a) and face grain perpendicular (b).

Table 4.5: Coefficient of friction for punched metal sheets and softwood.

		Face grain ∥		Face grain ⊥		End grain	
		μ_s	μ_k	μ_s	μ_k	μ_s	μ_k
Perforation 1	MEAN	-	-	0.82	0.49	-	-
	SD	-	-	0.13	0.09	-	-
	COV	-	-	16%	19%	-	-
Perforation 2	MEAN	0.51	0.42	0.48	0.43	0.49	0.42
	SD	0.07	0.03	0.05	0.04	0.03	0.04
	COV	13%	7%	10%	8%	6%	9%
Perforation 3	MEAN	0.80	0.61	0.67	0.62	0.88	0.74
	SD	0.08	0.08	0.08	0.09	0.06	0.07
	COV	10%	13%	11%	15%	7%	10%

4.3.5 Milled

Pyramid pattern

The test results for the DVW specimens with pyramid pattern continue at a high level, with average values of μ_s = 0.83 and 0.87 (0.5 mm), 0.89 and 0.95 (1.0 mm), 1.03 and 1.07 (1.5 mm) and 1.15 (2.0 mm) for face grain parallel and perpendicular respectively. Table 4.6 shows all results of μ_s and μ_k. The load-displacement curves followed diagram 1 (Figure 2.2a). However, the curve flattened out slightly before reaching the maximum value.

For the small pyramids with 0.5 mm depth, shear failure of the pyramids parallel to the loading direction occurred (Figure 4.13a). In addition, for some specimens with the smallest pyramids, local damage to the surface occurred due to knots (as was already observed with the DVW and embossed surface, see Figure 4.13b).

A strong interlocking effect was observed for the pyramids, leading to severe damage to the softwood specimens. This damage was especially severe for the tests with face grain perpendicular. The early wood, into which the pyramids were pressed, detached from the underlying late wood along the line of the annual rings (Figure 4.13c). This failure was differently pronounced depending on the annual rings' position but was not systematically investigated further. There was no strong interlocking for the tests with the face grain parallel to the loading direction; the pyramids could instead slide along the softwood surface (Figure 4.13d). Still, the difference in the friction coefficient for the two sliding directions is not as pronounced as the damaged softwood surfaces would suggest. All results, including the kinetic coefficients of friction, are given in Table 4.6.

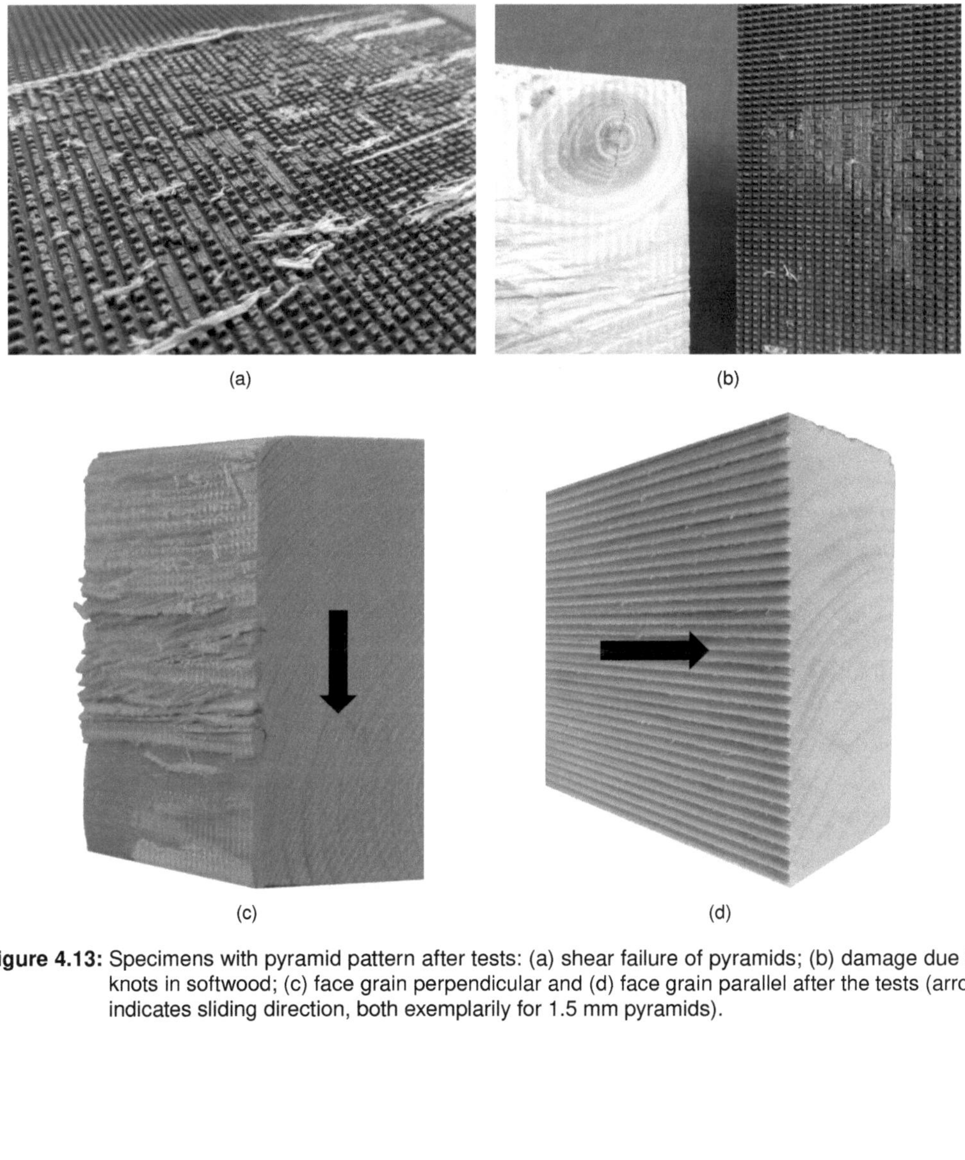

Figure 4.13: Specimens with pyramid pattern after tests: (a) shear failure of pyramids; (b) damage due to knots in softwood; (c) face grain perpendicular and (d) face grain parallel after the tests (arrow indicates sliding direction, both exemplarily for 1.5 mm pyramids).

Table 4.6: Coefficient of friction for milled pyramid patterns in DVW and softwood.

Depth of pyramids		Face grain ∥		Face grain ⊥		End grain	
		μ_s	μ_k	μ_s	μ_k	μ_s	μ_k
0.5 mm	MEAN	0.83	0.55	0.87	0.56	0.83	0.64
	SD	0.05	0.07	0.12	0.09	0.08	0.09
	COV	6%	13%	14%	16%	9%	14%
1.0 mm	MEAN	0.89	0.54	0.95	0.66	0.80	0.66
	SD	0.12	0.09	0.13	0.07	0.13	0.06
	COV	13%	16%	14%	10%	16%	8%
1.5 mm	MEAN	1.03	0.49	1.07	0.56	1.06	0.44
	SD	0.16	0.10	0.11	0.06	0.11	0.10
	COV	15%	21%	11%	11%	10%	23%
2.0 mm	MEAN	-	-	1.12	0.83	-	-
	SD	-	-	0.12	0.10	-	-
	COV	-	-	19%	13%	-	-

Circular pattern

The circular pattern was milled in both DVW and steel. As different cutting inserts for the milling tools were used, the surface geometries differed slightly. Also, the rotational speeds as well as feed rates differed, resulting in varying surface geometries. The mean values determined for the DVW and softwood were μ_s = 0.78 and 0.89 for face grain parallel and perpendicular, respectively. The curves followed mainly diagram 1 (Figure 2.2a). For the tests with end grain, the average value was slightly lower with μ_s = 0.82. After the tests, wood fibres sticking to the circular pattern were observed (Figure 4.14a), indicating some interlocking of the surface with the softwood. In general, the average values are close to the values of the 0.5 mm pyramid pattern, although the pattern was not milled as deep as the pyramids. The results confirm that the surface does not need prominent protruding features but rather evenly spread features that ensure tight contact with the adjoining surfaces.

For the tests with steel and softwood, the average values were μ_s = 0.58 and 0.63 for the tests with face grain parallel and perpendicular and μ_s = 0.64 for the tests with end grain. The curves also followed diagram 1 (Figure 2.2c). Because of the different cutting inserts, the results differ from the tests with DVW. No damage to the surface was observed, and the abrasion of fibres was less than for the tests with DVW (Figure 4.14b). All results, including the kinetic coefficients of friction, are given in Table 4.7.

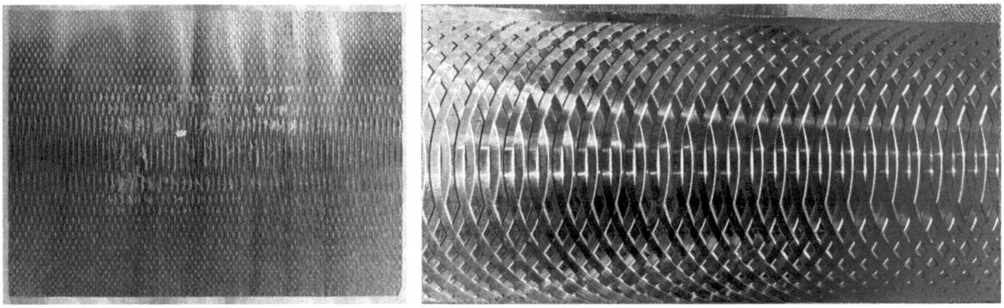

(a) (b)

Figure 4.14: Specimens with circular pattern in DVW (a) and steel (b) after the tests.

As seen in Table 4.7, the values for milled DVW are higher than for aluminium. That is because the tools used for the surface modification differed for the DVW and the steel specimens. Therefore, slightly different surfaces were created. The surface structure of the DVW was sharper, leading to a better interlocking of the surface and the softwood.

Table 4.7: Coefficient of friction for circular patterns (DVW and steel) and softwood.

		Face grain $\parallel$		Face grain $\perp$		End grain	
		μ_S	μ_K	μ_S	μ_K	μ_S	μ_K
DVW	MEAN	0.78	0.49	0.89	0.64	0.82	0.62
	SD	0.08	0.06	0.12	0.13	0.08	0.06
	COV	10%	12%	14%	20%	10%	9%
Steel	MEAN	0.58	0.42	0.63	0.46	0.64	0.52
	SD	0.05	0.03	0.05	0.07	0.11	0.09
	COV	9%	7%	8%	15%	17%	18%

Horizontal grooves

For the tests with steel and softwood, the average values were μ_S = 1.12 and 0.70 for the tests with face grain parallel and perpendicular and μ_S = 1.29 for the tests with end grain. No damage to the steel surface occurred. However, the friction coefficient was slightly higher for the first test with the freshly milled steel plate. This increase is assumed to result from chips remaining from the milling process. The curves followed mainly diagram 1 (Figure 2.2a); however, every time a horizontal groove met the softwood at the top of the specimens, the frictional force F_f increased again until sliding occurred. Figure 4.15 shows softwood specimens after the tests, which were severely damaged.

(a)

(b)

Figure 4.15: Softwood specimens after tests with horizontal grooves and face grain parallel (a) and face grain perpendicular (b).

Scale pattern

The scale pattern results from milling horizontal grooves and additional vertical grooves in DVW. On average, the friction coefficient was μ_s = 1.12 and 0.70 for the tests with softwood face grain parallel and perpendicular, and μ_s = 1.29 for the tests with end grain. The curves also followed diagram 1 (Figure 2.2a). No damage to the surface occurred, and the DVW specimens were used for multiple tests. Figure 4.16 shows the typical damage to the softwood specimens for the sliding direction parallel and perpendicular to the grain. All results, including the kinetic coefficients of friction, are given in Table 4.8.

(a)

(b)

Figure 4.16: Softwood specimens after tests with scale pattern and face grain parallel (a) and face grain perpendicular (b).

Table 4.8: Coefficient of friction for horizontal grooves (steel) or scale pattern (DVW) and softwood.

		Face grain ∥		Face grain ⊥		End grain	
		μ_s	μ_k	μ_s	μ_k	μ_s	μ_k
Horizontal grooves in steel	MEAN	1.12	0.69	0.70	0.55	1.29	0.96
	SD	0.13	0.07	0.06	0.05	0.16	0.20
	COV	11%	10%	9%	10%	12%	21%
Scale pattern in DVW	MEAN	0.40	0.29	0.61	0.48	0.58	0.44
	SD	0.09	0.04	0.10	0.15	0.04	0.02
	COV	22%	13%	16%	32%	6%	4%

4.3.6 Sanded

Tests with sanded DVW and softwood were only performed with face grain perpendicular and end grain. The mean values for the face grain perpendicular and end grain were calculated as μ_s = 0.56 and 0.47, respectively. The curves followed diagram 1 and 2 (Figure 2.2a+b), i.e. a clear peak of F_f and then pronounced stick-slip behaviour. No damage to the surface of the DVW was observed. The softwood showed minor sanding marks, and there was fine sawdust-like powder in the sanding grooves of the DVW; see Figure 4.17a. This observation did not occur in the tests with end grain.

4.3.7 Sandblasted

The tests with sandblasted DVW and softwood were also performed with face grain perpendicular and end grain only. The mean values for the face grain perpendicular and end grain were μ_s = 0.49 and 0.47, respectively. As with the sanded specimens, the curves followed diagram 1 and 2 (Figure 2.2a+b) with the clear peak and stick-slip afterwards. After the tests, fine sawdust could be detected on the surface. However, this did not adhere to the DVW and could be removed by lightly brushing over it. The sandblasted DVW specimen can be seen in Figure 4.17b.

(a) (b)

Figure 4.17: Specimens with (a) sanded and (b) sandblasted surface.

4.3.8 Brushed

Again, tests with brushed DVW and softwood were only performed with face grain perpendicular and end grain. The mean values for the face grain perpendicular and end grain were μ_s = 0.28 and 0.47, respectively. Although the surface seemed promising in terms of manufacture and surface feel, the results are relatively low. All results, including the kinetic coefficients of friction, are given in Table 4.9.

Table 4.9: Coefficient of friction for sanded, sandblasted, or brushed surface (all with DVW) and softwood.

		Face grain ⊥		End grain	
		μ_s	μ_k	μ_s	μ_k
Sanded DVW	MEAN	0.56	0.40	0.47	0.39
	SD	0.08	0.10	0.09	0.08
	COV	14%	25%	19%	21%
Sandblasted DVW	MEAN	0.49	0.41	0.47	0.44
	SD	0.07	0.05	0.06	0.05
	COV	15%	12%	13%	10%
Brushed DVW	MEAN	0.28	0.23	0.47	0.42
	SD	0.02	0.04	0.05	0.04
	COV	9%	17%	12%	10%

4.3.9 Coated

Flowable epoxy resin adhesive (2K EP)

The values determined for the coefficient of friction for tests with coated steel plates and softwood face grain parallel were μ_s = 0.75 for the coating with quartz sand 0.5–1.0 mm. For the tests with face grain perpendicular, the friction coefficients were μ_s = 0.68 for the sand 0.5–1.0 mm (coated steel plates), 0.64 for the sand 0–2 mm, and 0.61 for the grit (the later both coated DVW plates). For the tests with end grain, the results were similar with μ_s = 0.69, 0.54 and 0.69 for the coating with sand 0.5–1.0 mm, sand 0–2 mm and grit 2–4 mm respectively.

The coated steel specimens showed no adhesive failure. The sand grains stuck to the softwood specimens and were detached from the steel plate (Figure 4.18a). In contrast to this, mainly an adhesive failure of the coating was observed for the tests with the coated DVW, i.e. the adhesive detached from the DVW (Figure 4.18b+c). The used adhesive was intended for use with plastic, metal, and rubber materials. This unintentional use could explain the improper bond of the substrate to the surface of the DVW. All results, including the kinetic coefficients of friction, are given in Table 4.10.

Table 4.10: Coefficients of friction for coated surfaces (2K EP) and softwood.

		Face grain ∥		Face grain ⊥		End grain	
		μ_s	μ_k	μ_s	μ_k	μ_s	μ_k
Quartz sand 0.5–1.0 mm on steel	MEAN	0.75	0.66	0.68	0.61	0.69	0.65
	SD	0.07	0.04	0.04	0.06	0.13	0.08
	COV	9%	6%	6%	9%	19%	13%
Quartz sand 0–2 mm on DVW	MEAN	-	-	0.64	0.41	0.54	0.35
	SD	-	-	0.08	0.07	0.07	0.02
	COV	-	-	13%	17%	12%	7%
Grit 2–4 mm on DVW	MEAN	-	-	0.61	0.38	0.69	0.51
	SD	-	-	0.05	0.04	0.04	0.03
	COV	-	-	9%	10%	5%	6%

(a) (b) (c)

Figure 4.18: Coated specimens after tests: (a) steel and quartz sand 0.5–1.0 mm; (b) DVW and quartz sand 0–2 mm; (c) DVW and grit 2–4 mm.

Epoxy resin adhesive tape (EpoxyTape)

The results showed significantly higher coefficients of friction than were determined with the pasty epoxy resin. Thus, the tests with face grain resulted in mean values of μ_s = 0.82 for the 0.1 mm thick adhesive tape and μ_s = 0.74 for the 1.0 mm thick adhesive tape. The results of the tests with end grain showed even higher coefficients of friction with μ_S = 0.97 for the thin tape and μ_S = 0.82 for the thick tape.

Adhesive failure also occurred during the tests for the tape with an adhesive layer thickness of 0.1 mm. However, the interlocking of the quartz sand with the softwood was so strong that only isolated spots of the adhesive tape detached from the DVW (Figure 4.19a). Cohesive failure occurred in the tests with the tape with a layer thickness of 1.0 mm. The sand grains of the coating pressed into the softwood stuck there and were separated from the adhesive. This detachment of the sand grains from the adhesive tape can be seen in Figure 4.19b. During the tests with end grain, an exceptionally high degree of interlocking was observed, which was also observed with the notched surfaces. Figure 4.19c shows two softwood members with sand grains stuck to the end grain.

Table 4.11: Coefficients of friction for coated surfaces (EpoxyTape) and softwood.

		Face grain ⊥		End grain	
		μ_S	μ_K	μ_S	μ_K
Tape thickness 0.1 mm	MEAN	0.82	0.62	0.97	0.68
	SD	0.10	0.07	0.12	0.04
	COV	12%	11%	12%	6%
Tape thickness 1.0 mm	MEAN	0.74	0.50	0.82	0.61
	SD	0.03	0.12	0.05	0.00
	COV	4%	23%	6%	1%

(a) (b) (c)

Figure 4.19: Adhesive failure (a) and cohesive failure (b). Sand grains stuck to the softwood (c).

Grip tape

The surface of the grip tape was very rough, similar to sandpaper. However, it was not possible to determine an exact coefficient of friction during the tests, as adhesive failure also occurred with this type of coating. As seen in Figure 4.20a, the grip tape detached entirely from the DVW and shifted along the friction surface. Thus, coefficients of friction of only $\mu_S = 0.24$ and $\mu_S = 0.32$ could be determined for the tests with face grain perpendicular and end grain, respectively. Again, the tests with end grain showed high interlocking of the surfaces, with specimens sticking to the grip tape even after the tests were finished, see Figure 4.20b.

Table 4.12: Coefficient of friction for coated surface (grip tape) and softwood.

	Face grain ⊥		End grain	
	μ_S	μ_K	μ_S	μ_K
MEAN	0.24	0.10	0.32	0.10
SD	0.02	0.03	0.13	0.01
COV	9%	29%	40%	7%

(a)

(b)

Figure 4.20: Adhesive failure of the grip tape (a) and softwood specimens stuck to the grip tape (b).

4.4 Statistical evaluation

Probability distribution

To determine characteristic values (5% quantiles) of friction coefficients, EN 14358 [111] specifies statistical methods. The suggested parametric approach is only valid if the test data is normally or log-normally distributed. Therefore, in the first step, the probability distribution of the friction coefficient was determined, i.e. a log-normal distribution was assumed, and tests were performed to validate the assumption. This determination was exemplarily performed for the untreated and milled surfaces of the DVW specimens with logarithmic values of the friction coefficients. A Kolmogorov-Smirnov (KS) test was performed for both surfaces. The KS test is a statistical method to assess the similarity between two probability distributions. It is a non-parametric test, i.e. it makes no assumptions about the underlying distribution of the data. The test is based on the maximum difference between the cumulative distribution functions (CDF) of the two distributions being compared [78]. The test statistic k measures this difference and is used to determine whether the two distributions are significantly different from each other. The larger the test statistic, the more significant the difference between the two distributions being compared. The null hypothesis states that the data comes from a standard normal distribution (at the 5% significance level). For both tests, the test statistic k is less than the critical value c, indicating that the null hypothesis can be accepted ($0.0801 < 0.0954$ and $0.0445 < 0.0778$); see Figure 4.21. That the test fails to reject the null hypothesis is also confirmed by the parameter h, which is either 0 (failure to reject the null hypothesis at the significance level) or 1 (rejection of the null hypothesis at the significance level). The returned p-values are well above the significance level of 0.05 ($p = 0.1471$ and 0.5782). The KS test results suggest a log-normal distribution for both surfaces. This suggestion is confirmed by the histograms with the bell curve for the normal distribution and the quantile-quantile plots (all points close to identity line $x = y$), see Figure 4.21.

A log-normal distribution was confirmed for the results of the tests with the most specimens, i.e. untreated DVW and milled DVW. Therefore, it was assumed that all other test series would also be log-normally distributed without further testing the null hypothesis.

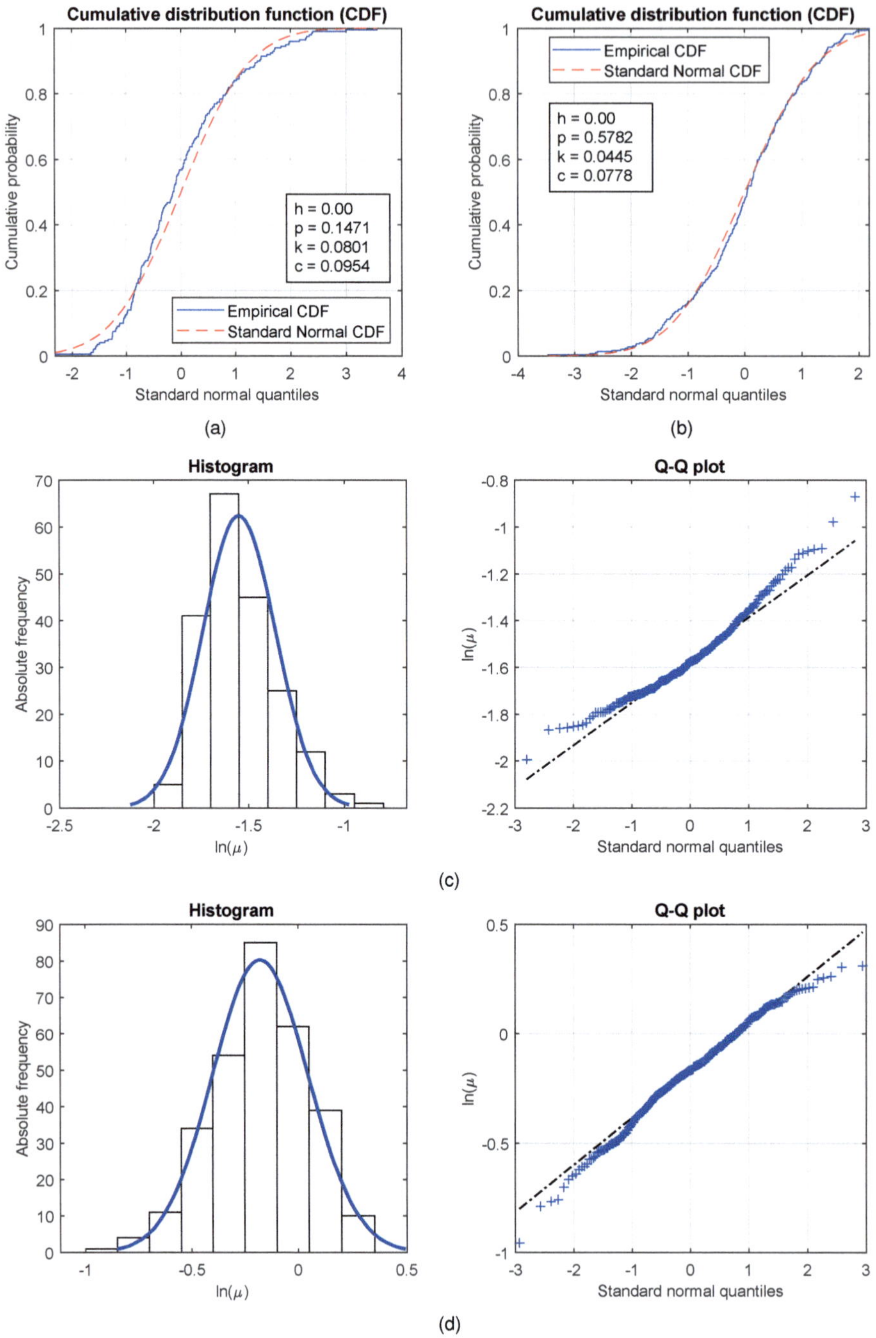

Figure 4.21: Results of Kolmogorov-Smirnov test: (a) untreated surface; (b) milled surface. Log-normal distribution of friction coefficients: (c) untreated surface; (d) milled surface.

Analysis of variance (ANOVA)

An analysis of variance (ANOVA) was performed with SAS® analytics software [120] to test if there is a statistical difference between the mean values of the friction coefficients of the different surfaces and especially of the different grain directions. This method applies when the random samples are independent, the population distributions are normal, and all exhibit the same variance. After establishing the null hypothesis, assuming equal means, and the alternative hypothesis, positing at least one difference among them, a predetermined significance level is set [78]. Here, a one-way ANOVA was used, checking the means of three independent groups with only one variable (or "factor") at a significance level of 5%. The null hypothesis is either rejected or not. The null hypothesis states that all group means are equal. If the p-value exceeds 0.05, the test fails to reject the null hypothesis. There is not enough evidence to say that there is a significant difference between the groups. If no actual variance (no real difference) exists between the groups, the F-ratio should equal close to 1. If the p-value is smaller than 0.05, the null hypothesis can be rejected, and there is a significant difference between the groups. The larger the F-value, the greater the evidence of a difference between the groups. A post hoc test was conducted to see which groups were different. Here, Tukey's HSD Test for multiple comparisons was used.

The results of the ANOVA for the untreated DVW surface are already given in the parameter study performed in Chapter 2, Section 2.5. Here, the results of the performed ANOVA for the milled surfaces are given. This surface was the surface modification with the most tests and, therefore, with the greatest significance.

Surfaces with no statistically significant differences:

For the tests with the 0.5 mm pyramid pattern, there is no significant difference between the three grain directions ($p = 0.2022$ and $F = 1.63$). Also, for the tests with the 1.5 mm pyramid pattern, there is no significant difference between the three grain directions ($p = 0.6292$ and $F = 0.47$). Finally, for the tests with the circular pattern milled in steel, there is no significant difference between the three grain directions ($p = 0.1221$ and $F = 2.20$).

Surfaces with statistically significant differences:

For the tests with the 1.0 mm pyramid pattern, there is a significant difference between the three grain directions ($p = 0.0295$ and $F = 3.83$). The Tukey post hoc test showed a significant difference between the tests with face grain perpendicular and end grain; however, there was no difference between the other grain directions. The tests with the circular milled pattern in DVW showed a significant difference between the grain directions ($p = 0.0083$ and $F = 5.36$). The Tukey post hoc test showed a significant difference between the tests with face grain perpendicular and face grain parallel; however, there was no difference between the other grain directions. Additionally, the embossed surface with the inverse pyramid pattern was tested. The tests showed a significant difference between the three grain directions ($p = 0.0005$ and $F = 8.72$). The Tukey post hoc test showed a significant difference between the tests with face grain perpendicular and end grain and between face grain perpendicular and face grain parallel. No difference between the face grain parallel and the end grain was determined.

In addition, the milled surfaces (only face grain perpendicular) were compared with a one-way ANOVA. As expected, there is a statistically significant difference between the different milled surfaces ($p = 0.0001$ and $F = 907.11$). The Tukey's HSD Test for multiple comparisons showed that all surfaces are different, except for the milled pyramids with 0.5 mm and 1.0 mm grooves and the pyramids with 1.5 mm and 2.0 mm grooves. That is quite interesting as it shows that the extra time it takes to manufacture the smaller pyramids (i.e. 0.5 mm and 1.5 mm) can be saved, as there is no difference with the next larger pyramid size. Conversely, that also means that the additional depth of the larger pyramids (i.e. 1.0 mm and 2.0 mm) does not necessarily lead to greater friction coefficients.

Finally, the Tukey test showed no difference between the milled 0.5 pyramid pattern and the milled circular pattern. With both surface modifications, the milling head does not penetrate deeply into the surface, resulting in a similar surface roughness. The result of the ANOVA implies that the circular pattern can be milled as an alternative to the complex milling of the pyramid pattern.

4.5 Characteristic values

The characteristic values (5% quantiles) of friction coefficients were determined according to EN 14358. The suggested approach is only valid if the test data is normally or log-normally distributed. As the previously performed KS tests indicated a log-normal distribution of the friction coefficient, the calculation of the characteristic values was carried out according to Equation 4.2:

$$m_{\mathrm{k}} = \exp(\overline{y} - k_{\mathrm{s}}(n) s_{\mathrm{y}}) \tag{4.2}$$

with m_k characteristic value of the sample
$\bar{y}$ mean value
$k_s(n)$ statistical value according to Equation 4.3
s_y standard deviation

The following simplified expression may be used to determine $k_s(n)$:

$$k_s(n) = \frac{6,5n + 6}{3,7n - 3} \tag{4.3}$$

According to EN 14545 [112], the characteristic value may be calculated using a global coefficient of variation COV_g based on all test results n. Likewise, $k_s(n)$ may be calculated with the same number n of results. COV_g is calculated according to Equation 4.4:

$$COV_g = \sqrt{\frac{\sum((n_i - 1)(s_{y,i})^2)}{(\sum n_i - J)}} \geq 0.05 \tag{4.4}$$

with n_i number of test results per surface and fibre direction
$s_{y,i}$ standard deviation per surface and fibre direction
J number of test series

A total of n = 844 friction tests were evaluated. The global coefficient of variation was calculated according to Equation 4.4 to COV_g = 0.16 for the static friction coefficients and 0.15 for the kinetic friction coefficients. The statistical value according to Equation 4.3 was $k_s(n)$ = 1.76. In Table 4.13, the mean and corresponding characteristic values for the static friction coefficients are given for all investigated surfaces and grain directions. Table 4.14 respectively shows the values for the kinetic friction coefficients. The 5% quantile values for most surface modifications exceed the value of μ = 0.25 from Eurocode 5.

Table 4.13: All determined static coefficients of friction: Mean and characteristic values.

No.	Surface	Static coefficient of friction μ_s					
		FG ∥		FG ⊥		EG	
		mean	char.	mean	char.	mean	char.
0.1	Untreated DVW	0.22	0.17	0.23	0.17	0.20	0.15
0.2	Untreated aluminium	0.25	0.19	0.34	0.25	0.39	0.29
1.1	File cut (steel)	1.83	1.36	1.06	0.80	1.11	0.82
1.2	Rasp cut (steel)	2.60	1.74	1.33	0.85	1.47	1.11
2.1	Negative pyramid pattern (DVW)	0.68	0.51	0.79	0.59	0.71	0.53
2.2	Chequered plate (steel)	0.66	0.48	0.59	0.43	0.58	0.44
3.1	Perforation 1 (steel)	-	-	0.82	0.61	-	-
3.2	Perforation 2 (steel)	0.51	0.38	0.48	0.36	0.49	0.37
3.3	Perforation 3 (steel)	0.80	0.60	0.67	0.50	0.88	0.67
4.1	Pyramid pattern						
4.1.1	0.5 mm (DVW)	0.82	0.62	0.87	0.65	0.83	0.62
4.1.2	1.0 mm (DVW)	0.89	0.67	0.95	0.71	0.80	0.60
4.1.3	1.5 mm (DVW)	1.03	0.77	1.07	0.80	1.06	0.80
4.1.4	2.0 mm (DVW)	-	-	1.12	0.83	-	-
4.2	Circular pattern						
4.2.1	DVW	0.78	0.59	0.89	0.66	0.82	0.62
4.2.2	Steel	0.58	0.44	0.63	0.48	0.64	0.47
4.3	Horizontal grooves (steel)	1.12	0.84	0.70	0.52	1.29	0.97
4.4	Scale pattern (DVW)	0.40	0.30	0.61	0.45	0.58	0.44
5	Belt grinding (DVW)	-	-	0.56	0.42	0.47	0.35
6	Sandblasting (DVW)	-	-	0.49	0.36	0.47	0.35
7	Brushing (DVW)	-	-	0.28	0.21	0.47	0.35
8.1	2K-SE						
8.1.1	Sand 0.5–1.0 mm (steel)	0.75	0.57	0.68	0.51	0.69	0.51
8.1.2	Sand 0–2 mm (DVW)	-	-	0.64	0.48	0.54	0.41
8.1.2	Grit 2–4 mm (DVW)	-	-	0.61	0.46	0.69	0.52
8.2	EpoxyTape						
8.2.1	Tape 0.1 mm (DVW)	-	-	0.82	0.62	0.97	0.73
8.2.2	Tape 1.0 mm (DVW)	-	-	0.74	0.56	0.82	0.61
8.3	Griptape (DVW)	-	-	0.24	0.18	0.32	0.22

FG = face grain and EG = end grain

Table 4.14: All determined kinetic coefficients of friction: Mean and characteristic values.

No.	Surface	Kinetic coefficient of friction μ_k					
		FG $\parallel$		FG $\perp$		EG	
		mean	char.	mean	char.	mean	char.
0.1	Untreated DVW	0.19	0.15	0.19	0.14	0.19	0.14
0.2	Untreated aluminium	0.24	0.18	0.41	0.30	0.45	0.34
1.1	File cut (steel)	0.60	0.45	0.61	0.46	0.86	0.64
1.2	Rasp cut (steel)	0.69	0.48	0.74	0.54	1.37	1.02
2.1	Negative pyramid pattern (DVW)	0.50	0.38	0.52	0.38	0.52	0.39
2.2	Chequered plate (steel)	0.65	0.48	0.61	0.44	0.64	0.48
3.1	Perforation 1 (steel)	-	-	0.49	0.37	-	-
3.2	Perforation 2 (steel)	0.42	0.32	0.43	0.32	0.42	0.32
3.3	Perforation 3 (steel)	0.61	0.45	0.62	0.46	0.74	0.67
4.1	Pyramid pattern						
4.1.1	0.5 mm (DVW)	0.55	0.41	0.56	0.42	0.64	0.48
4.1.2	1.0 mm (DVW)	0.54	0.40	0.66	0.50	0.66	0.50
4.1.3	1.5 mm (DVW)	0.49	0.36	0.56	0.42	0.44	0.33
4.1.4	2.0 mm (DVW)	-	-	0.83	0.62	-	-
4.2	Circular pattern						
4.2.1	DVW	0.49	0.37	0.64	0.47	0.62	0.46
4.2.2	Steel	0.42	0.31	0.46	0.35	0.53	0.38
4.3	Horizontal grooves (steel)	0.69	0.52	0.55	0.41	0.96	0.71
4.4	Scale pattern (DVW)	0.29	0.22	0.48	0.35	0.44	0.33
5	Belt grinding (DVW)	-	-	0.40	0.29	0.39	0.29
6	Sandblasting (DVW)	-	-	0.41	0.30	0.44	0.32
7	Brushing (DVW)	-	-	0.23	0.17	0.42	0.31
8.1	2K-SE						
8.1.1	Sand 0.5–1.0 mm (steel)	0.66	0.50	0.61	0.46	0.65	0.49
8.1.2	Sand 0–2 mm (DVW)	-	-	0.41	0.31	0.35	0.26
8.1.2	Grit 2–4 mm (DVW)	-	-	0.38	0.28	0.51	0.38
8.2	EpoxyTape						
8.2.1	Tape 0.1 mm (DVW)	-	-	0.62	0.47	0.68	0.51
8.2.2	Tape 1.0 mm (DVW)	-	-	0.50	0.37	0.61	0.46
8.3	Griptape (DVW)	-	-	0.10	0.08	0.10	0.07

FG = face grain and EG = end grain

4.6 Conclusion

A total of 26 different surfaces, all of which were manufactured differently, were investigated. Most surface treatments showed a significant increase in the coefficient of friction compared to the untreated surface and the aluminium surface of conventional system connectors.

For reference values, tests with untreated DVW and aluminium were carried out. The results of the tests with aluminium (average values of all tests, disregarding grain direction) match well with values from literature (i.e. μ_s = 0.35 and μ_k = 0.23).

The surface modification of notching (file cut and rasp cut) is by far the most effective. Especially for tests with rasp cut and face grain, the effect is close to an interlocking, as seen with punched metal plate fasteners. Therefore, the term "friction" should be used with caution. Nevertheless, further efforts should be made to realise this type of surface modification for larger, flatter steel components. As long as the deformations in the shear plane are kept small and no sliding occurs, this surface offers the most significant potential for very stiff and strong connections.

Another very promising surface modification is the embossing of the pyramid pattern. In contrast to the notched surfaces, the results for the friction coefficient are only half of that. Nevertheless, the embossed surface is preferred to the other surfaces, especially considering the manufacturing process. Also, the friction coefficients are almost similar in all three sliding directions. The tests with this surface showed that a uniform surface directly in contact with its counterpart is more critical than protruding features that must be pressed into its counterpart.

Like the embossed surface, the performance of the circular milled pattern should be mentioned. The determined friction coefficients are higher than for the embossed surface, and the manufacturing process is also simple and very effective. These tests confirmed the findings about a uniform surface. However, the surface features depend on the size of the milling cutter and the inserts. Here, further efforts must be made to optimise the milling inserts to achieve a surface with "sharper" features.

The results with the different pyramid patterns are auspicious, as characteristic values of the coefficient of friction with softwood of more than twice the current value in EC 5 were determined. The most significant advantage of this surface modification is the identical surface of a connector in all possible loading directions. The manufacturing process guarantees a uniform surface that is easily replicable. However, the manufacturing process is also the most significant disadvantage because, until now, each groove must be milled individually. Further efforts are needed to resolve this tedious manufacturing process. Nevertheless, this challenge should be solvable in an industrial application. With steel/aluminium parts, there is also no risk of the small pyramids shearing off.

The tests with the coated specimens showed that the thinnest possible adhesive layer thicknesses are recommended for coating. Likewise, mineral grain sizes that are as fine as possible should be preferred. Here, further tests should be carried out with a grain size $\leq$ 0.5–1.0 mm, such as a coating with corundum. However, a couple of unknowns with coated surfaces arise. On the one hand, the durability of the adhesive must be guaranteed to prevent adhesive failure. However, as there are already approved adhesive systems for timber structures (e.g. glued-in threaded rods), this point should not be hard to solve.

On the other hand, it must be ensured that the coating used does not contain any oversized particles and that there are, therefore, individual "peaks" over which the entire friction surface tilts (see also the notes in [8] and [9] on increasing friction with sand). However, this problem should also be solvable or not occur in the first place through quality control during the manufacturing process. The last point concerns the storage, transport, and handling of possible coated connectors. It must be ensured that the coating is not (partially) abraded before installation.

In order to classify the test results of the friction tests against the background of the manufacturing of the surfaces, a (subjective) assessment of the surfaces was carried out. Table 4.15 gives an overview of the investigated surface modifications and the complexity of their manufacturing process. The classification reflects a balance between practicality, customisation, and efficiency in achieving the desired surface characteristics. The grading of the coefficient of friction refers to an average coefficient of friction for softwood on softwood of $\mu = 0.49$ (see Chapter 2, Table 2.1). Therefore, surfaces with $\mu < 0.50$ are marked with – (and for $\mu < 0.40$ with – –). Surfaces with $\mu > 0.50$ are marked with + (and for $\mu > 0.75$ with + + and for $\mu > 1.0$ with + + +).

Table 4.15: Advantages and disadvantages of the individual surface modifications.

Surface	Manufacturing process	Coefficient of friction
Notching		
File cut	–	+ + +
Rasp cut	–	+ + +
Embossing		
Inverse pyramid pattern	+ +	+ +

Continued on the next page

Table 4.15 – continued from previous page

Surface	Manufacturing process	Coefficient of friction
Chequered plate	+ +	+
Punching		
Perforation 1	+ +	+ +
Perforation 2	+ +	–
Perforation 3	+ +	+
Profile milling		
Pyramid pattern		
0.5 mm	– –	+ +
1.0 mm	– –	+ +
1.5 mm	– –	+ + +
2.0 mm	– –	+ + +
Circular pattern		
DVW	+	+ +
Steel	+	+
Horizontal pattern	+ +	+
Scale pattern	+	+
Belt grinding	+ +	+
Sandblasting	+ +	–
Brushing	+ +	– –
Coating		
2K-SE		
Sand 0.5–1.0 mm	– –	+
Sand 0–2 mm	– –	+
Grit 2–4 mm	– –	+
EpoxyTape		
Tape 0.1 mm	–	+ +
Tape 1.0 mm	+	+
Grip tape	+ +	– –

For the following connection tests, mainly milled surfaces (pyramid pattern, circular pattern) and mainly "flat" surfaces (embossed pattern, circular pattern) with no (extensive) protruding features are further investigated.

5 Connection tests

5.1 Introduction

Connections in timber engineering with mechanical fasteners are state-of-the-art and widely used. Dowel-type fasteners are either inserted perpendicular or inclined to the shear plane. Fasteners inserted perpendicular are mainly laterally loaded. Their load-carrying capacity is limited by the timber's embedment strength and the fastener's bending strength. Due to plastic deformation of the fasteners at large displacements, these types of connections exhibit ductile behaviour (if the connection is designed accordingly with plastic hinges). Fasteners inserted inclined at an angle to the shear plane are mainly axially loaded. Their load-carrying capacity is limited by their withdrawal strength and their tensile strength. These connection types fail brittle (if the connection is designed accordingly, whereby the tensile strength is decisive), but have a high longitudinal stiffness with small displacements. For both types of connections, friction in the shear plane is activated with increasing deformation. In the following chapter, tests were performed with fasteners inserted perpendicular or inclined to the shear plane and modified surfaces to increase the friction in the shear plane. Some results have already been published in [3].

5.2 Connections with laterally loaded fasteners

5.2.1 Test programme and setup

Double-shear steel-to-timber connections with bolts as fasteners were tested. In the first series, untreated steel plates were used to create reference values. In subsequent test series, steel plates with a milled circular pattern and steel plates coated with quartz sand were used. The surface modification of coating was chosen, as good experience with coating steel with quartz sand was gained in previous research [79]. The test programme is given in Table 5.1. The tests investigated whether a sufficiently high contact pressure can be generated in the shear plane with bolts inserted perpendicular to the shear plane and whether the previously determined friction coefficients occur. Some tests were performed by Wursthorn [80] as part of his Bachelor's Thesis.

Table 5.1: Test programme for tests with laterally loaded bolts and different surface modifications.

Series	No. of tests	Surface	Bolt type and count	
1	3	Untreated	10x240	1
2	3	Milled	10x240	1
3	5	Coated	10x240	1

The experimental setup can be seen in Figure 5.1a. The dimensions of the test specimens are given in Figure 5.1b. The specimens were designed so that failure with two plastic hinges per shear plane is decisive. By mistake, the end distance $a_{3,t}$ was only $7d$ (instead of $8d$ according to EC 5). The steel surface was modified on both sides of the plates. The bolts used for all tests were of strength class 4.6 and had a diameter of d = 10 mm and a length of ℓ = 240 mm. Washers with d = 50 mm were used on both sides of the bolts. The nuts were tightened with a torque-controlled spanner with a torque of 25 Nm. The axial force of the bolts was not further controlled or measured. The test specimens were loaded with a universal testing machine.

The machine load and the relative displacement of each shear plane were measured during the test procedure at a rate of 1 Hz. The relative displacement was measured at each connection at the front and back of the test specimens. Hereby, the displacement evaluated was the relative displacement of the steel part to the timber side part. For all test series, the displacement was measured with the help of a digital image correlation (DIC) system (LIMESS Messtechnik u. Software GmbH). The location of the markers used to determine the displacement can be seen in Figure 5.1a. All markers were placed on the neutral axis of the fasteners and in the centre of each part. The resolution of the cameras was 4112 x 2176 P (= 8.9 MP). The field of view was chosen to approximately 400 x 755 mm^2. The point size of the speckle pattern was approximately 10 P (= 1 mm). According to the supplier, the facet size was set to 31 P and the grid space to 9 P. The test procedure and the evaluation were carried out following EN 26891 [116] with an unloading loop in the range of 10–40% of F_{est}. Both the maximum load $F_{V,test}$ and the stiffness k_s were determined per connector and shear plane. To identify the surface's influence on the stiffness, the initial stiffness k_i was determined.

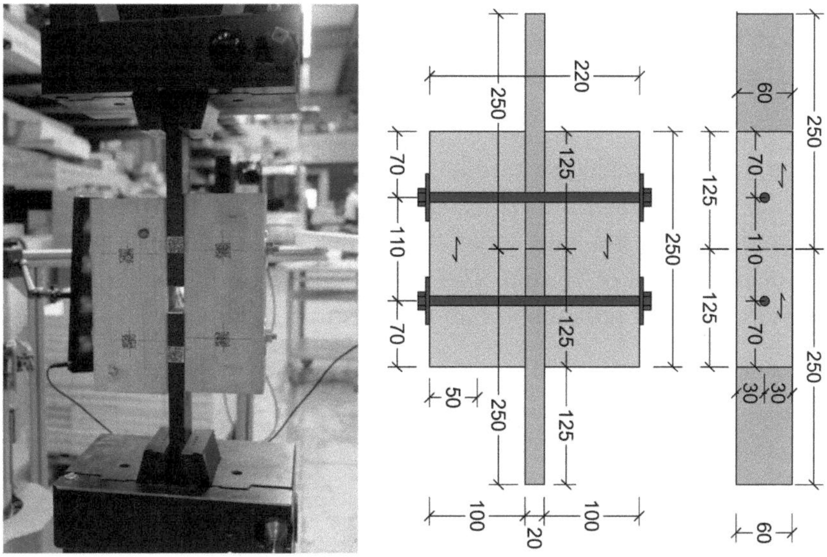

(a) (b)

Figure 5.1: Test setup for tests with perpendicular bolts (a) and measurements of test specimens (b).

5.2.2 Results and discussion

The ultimate loads and the stiffness values are given per fastener and shear plane. The ultimate load is either the load at failure of the specimen or at a displacement of 15 mm. The load range the stiffness values were determined for, changed individually and is, therefore, always explicitly given.

Series 1: Untreated surface

The tests with the untreated steel surface reached a mean ultimate load of $F_{V,test}$ = 14.8 $\pm$1.9 kN. Figure 5.5a shows the load-displacement curves for all tests. The load increases right at the start of the test to approximately 4 kN without significant displacement. This increase is due to the tightening of the nut and, therefore, prestressing the shear plane, which activates friction between steel and timber. After the frictional force is overcome, the load increases linearly to approximately a displacement of 2 mm. At around this point, the stiffness decreases, and the curve flattens. Splitting occurs in the timber members, and the load slightly decreases before the tests are terminated.

The stiffness k_s was determined in the range of 5.5–9.0 kN, which is about 35–60% of $F_{V,test}$. The stiffness k_s averages to 3.87±0.5 kN/mm. Additionally, the stiffness k_i of the very first section of the load-displacement curve was determined in the range of 0–3.2 kN (ca. 0–22%) and averages to 110±34 kN/mm.

As designed, two plastic hinges form in the bolt, as shown in Figure 5.2a. As shown in Figure 5.2b, the timber starts splitting with increasing load. After the first cracks appeared in the timber, the load could still be increased; however, none of the tests reached a displacement of 15 mm. The specimens were not reinforced with screws against splitting, as this precaution had not been considered during the preparation process.

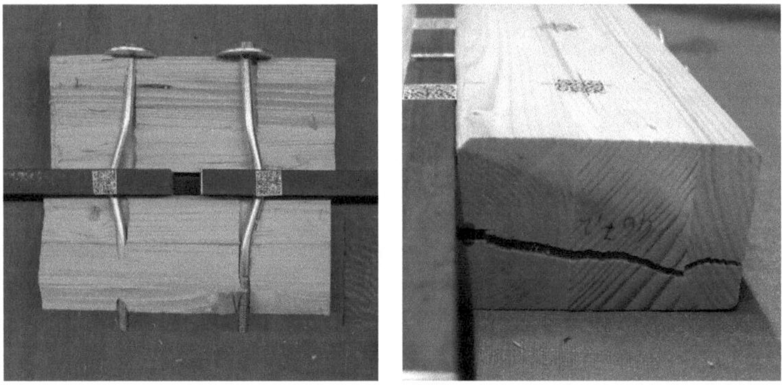

(a) (b)

Figure 5.2: Plastic hinges in bolts (a) and splitting of timber member (b) [80].

Series 2: Milled surface

The tests with the milled steel surface showed similar behaviour to the tests before. A higher mean ultimate load of $F_{V,test}$ = 17.1±0.6 kN could be reached in the tests. Figure 5.5b shows the load-displacement curves for all tests. The load increases initially to approximately 5 kN, before horizontal slip occurs. After exceeding the static frictional force, the load rises more or less linearly and up to approximately 2 mm displacement before the stiffness decreases and the curve flattens. Afterwards, the load can be constantly increased to a displacement of 15 mm, where the test was terminated.

The stiffness k_s was determined in the range of 6.5–10 kN, which is also about 35–60% of $F_{V,test}$. Interestingly, the stiffness k_s is slightly lower than before and averages 3.65±0.4 kN/mm. However, the stiffness k_i is significantly higher and averages to 152±30 kN/mm and was determined in the range of 0–3.8 kN (ca. 0–22%).

As well as before, two plastic hinges form in the bolt. As shown in Figure 5.3a, no damage to the timber surface occurs and barely any abrasion can be detected. The timber starts splitting with increasing load, as shown in Figure 5.3b. However, the cracks were minor and did not necessarily go through the whole timber member. Therefore, it is understandable that no failure prior to a displacement of 15 mm occurred.

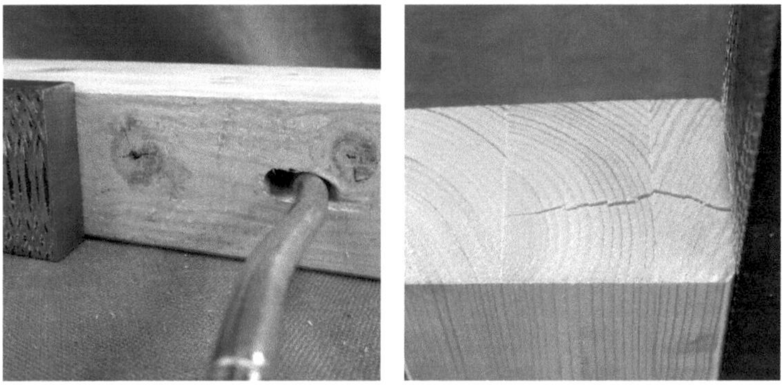

(a) (b)

Figure 5.3: Undamaged timber surface beneath milled steel surface (a) and splitting of timber member (b).

Series 3: Coated surface

The final tests with the coated steel surface again show similar behaviour to the other tested surfaces. The mean ultimate load can be increased again to $F_{V,test}$ = 18.3±2.6 kN. Figure 5.5c shows the load-displacement curves for all tests. However, the initial load increase due to the static frictional forces is the lowest, and after approximately 3 kN, the frictional force is exceeded, and horizontal displacement is recorded. The load rises linearly to approximately 2 mm displacement before the curve flattens. After a total displacement of about 6 mm, the stiffness increases slightly, and the gradient of the curves rises. Splitting of the timber occurred, but the load did not drop before reaching a displacement of 15 mm (except for one specimen).

The stiffness k_s was also determined in the range of 6.5–10 kN, which again is about 35–60% of $F_{V,test}$. The stiffness k_s is the highest of all tested surfaces and averages

3.96±0.5 kN/mm. However, the stiffness k_i is significantly lower than before and averages only 72±15 kN/mm. The stiffness was determined in the range of 0–2.5 kN (ca. 0–14%).

Figure 5.4a shows the coated surface after the tests. No evident damage to the surface is visible. However, many abrasions of timber fibres sticking to sand grains can be noted. This abrasion can also be seen in Figure 5.4b at the end grain end of the timber member. Here, the splitting of the timber is also visible.

(a) (b)

Figure 5.4: Abraded timber surface beneath coated steel surface (a) and at the end grain end of timber member (b).

Comparison

All test results are given in Table 5.2. Figure 5.5 and Figure 5.6 show the load-displacement curves for all steel-to-timber tests. Coloured is always one test per series. Overall, the load-displacement behaviour for all three different surfaces is very similar. Particularly in the range between 5–10 kN, the curves are almost congruent, which is also reflected in the stiffness values of k_s. Between 2–3 mm, all tests show an inflexion point with a subsequent flatter gradient. Significant surface differences can be observed directly at the start of the test. As expected, adhesion can be increased by a higher coefficient of friction in the tests with a milled surface. However, the tests with the coated surface do not show such a high level of adhesion despite the higher coefficient of friction. This reduced adhesion is due to the surface type and is consistent with observations from the friction tests. The grains of sand cannot be pressed sufficiently into the timber under the specified contact pressure of the bolts. As a result, there is no interlocking of the surfaces but rather a “rolling” of the timber surface over the coating.

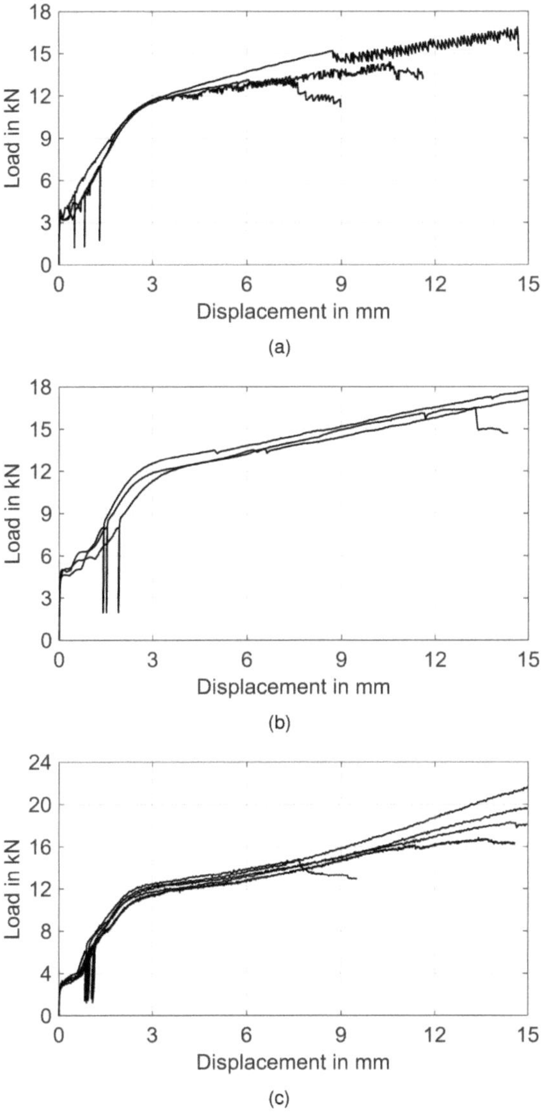

Figure 5.5: Force-displacement curves for all steel-to-timber tests: (a) untreated steel plate; (b) milled steel plate; (c) coated steel plate.

With greater deformations, the three surfaces also differ from each other. The wood splits in the tests with untreated steel, and an associated load drop occurs. The test specimens with the milled surface also split due to the large deformations and the associated inclination of the fastener; the rope effect is activated, and there is an additional force due to friction, which increases the overall load-carrying capacity. For both surfaces, untreated and milled, the stiffness after the inflexion point is still linear, albeit with a smaller gradient. However, in the tests with the coated surface, the stiffness increases again, and the curve gradient increases. This increased gradient is because, with correspondingly large deformations, the contact pressure in the shear plane increases and the sand grains are now entirely pressed into the timber. The resulting increase in stiffness and load-carrying capacity is recognisable.

Table 5.2: Ultimate loads and stiffness values from the tests with bolts and different surfaces.

Series	Surface	$F_{V,test}$ in kN	k_i in kN/mm	k_s in kN/mm
1	Smooth	14.8±1.9	110±34	3.87±0.5
2	Milled circular	17.1±0.6	152±30	3.65±0.4
3	Coated with quartz sand	18.3±2.6	72±15	3.96±0.5

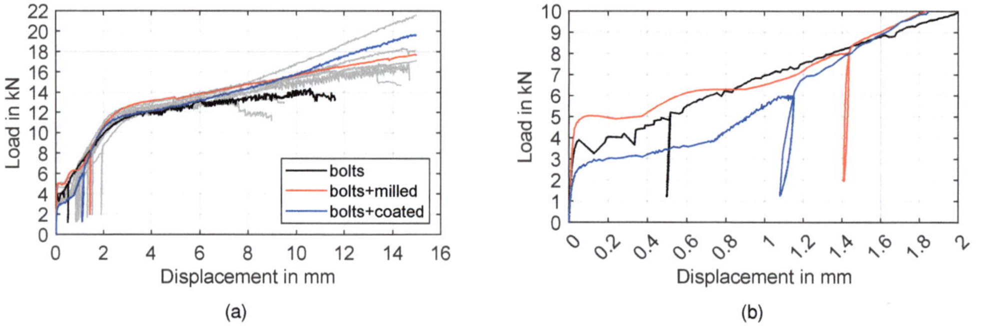

Figure 5.6: Load-displacement curves for tests with laterally load fasteners: (a) total displacement; (b) detail of initial stiffness.

5.2.3 Summary

Steel-to-timber tests were performed with fasteners inserted perpendicular to the shear plane. Under loading parallel to the shear plane, these fasteners were laterally loaded. Additionally, the surface of the steel plates was modified to increase the friction in the shear plane. The surface modifications were a milled circular pattern and a coating with quartz sand. The results showed an increase in load-carrying capacity for both modified surfaces. The stiffness in the range of 35–60% of the ultimate load was similar for all three surfaces. An increased initial stiffness was observed in the range of 0–20%. However, the frictional force that had to be exceeded was not due to the fastener's inclination and the rope effect but rather to the nuts' tightening prior to the tests (here 25 Nm). This is also the only way to guarantee the initial stiffness, if the bolts are all equally tightened with a specified torque.

In conclusion, an increased coefficient of friction also leads to an increased load-carrying capacity. However, a large displacement is needed to exert the appropriate contact pressure. Surfaces with protruding parts, such as the coated surface, are disadvantageous initially, as the surfaces tend to slide against each other rather than interlock. Especially at the beginning of the tests, the preload of the bolts can influence the stiffness to a much greater degree than the surface modification.

5.3 Connections with inclined fasteners

5.3.1 Test programme

In the first series of tests, simple connectors made of densified veneer wood (DVW) with a modified surface were used. The load between the two connector parts was transmitted via compressive contact. The first aim was to investigate whether a sufficiently high contact pressure can be generated in the shear plane with inclined screws in a later installation situation and, thus, whether the previously determined friction coefficients occur. Furthermore, the extent to which the increase in the coefficient of friction of the surface affects the increase in the load-carrying capacity of the connection was investigated. The tests were performed by El Hamoui [81] as part of his Bachelor's Thesis.

In a second series of tests, the position of the screws was changed, and only two milled surfaces were examined. However, the type and number of screws remained the same. Subsequently, in the third series of tests, the screw type was changed, and longer screws were selected so that the tensile capacity of the screws became decisive. Again, only two milled samples were examined. With the increase in the number of screws in the fourth series of tests, the behaviour of the connectors with a larger group of screws was examined. The aim was to check whether, with the large number of screws arranged very close to each other, block shear failure must be taken into account in the subsequent design model. In a further fifth series of tests, tests were also carried out with longer screws.

In Series 6, two prototype connectors made of DVW with milled surfaces were tested for the first time. The prototypes v1 and v2 were fastened with 10 and 12 fully threaded screws 6x200 mm, which were inclined by 45° to the shear plane. The surface was modified with a milled pyramid pattern, i.e. pyramids with a depth of 0.5 mm for prototype v1 and a depth of 1.0 mm for prototype v2. Additionally to the inclined screws, the prototyp connector v2 was fastened to the timber members with a total of four fully-threaded screws (in the following named as assembly screws) that were inserted perpendicular to the shear plane. In Series 7, a heavy-duty prototype connector for forces up to 500 kN was tested. This prototype v3 was fastened with 20 screws 8x300 mm. Table 5.3 gives an overview of all performed push-out tests. The number of screws is given per connector plate and shear plane.

5.3.2 Test setup

The experimental setup can be seen in Figure 5.7. The exact dimensions of the test specimens are given in Appendix A.2, Figure A.2 and Figure A.6. The DVW connectors were modified on one side only, i.e. the side of the shear plane, and fastened to the softwood side and middle members with inclined screws. The screws used for Series 1 and 2 were fully threaded HECO-TOPIX [118] screws with a diameter of d = 5 mm. For the Series 3 to 7, fully threaded Würth ASSYplus VG 4 [117] screws with a diameter of d = 6–8 mm were used. The screws were inserted at an angle of 45° for all tests. The test specimens were loaded with a universal testing machine. During the test, both the machine load and the relative displacement between the middle and side members were measured. The relative displacement was measured at each connector at the front and back of the test specimens. For the first test series, inductive displacement transducers were used (measuring rate of 5 Hz); for all other test series, the deformation was measured (measuring rate of 1 Hz) with the help of a digital image correlation (DIC) system (LIMESS Messtechnik u. Software GmbH). The location of the markers (and the LVDTs) used to determine the displacement can be seen in Figure 5.7. All markers/LVDTs were placed on the timber parts at approximately half the height of the respective connector plate. The resolution of the cameras was 4112 x 2176 P (= 8.9 MP). The field of view was chosen differently for all test series and not further documented. The speckle patter for the tests of series 2–7 was spray painted and, therefore, contained various point sizes. According to the supplier, the

Table 5.3: Overview of the push-out tests.

Series	No. of tests	Surface	Screw type and number	
1	3	Untreated	5x100	5
	3	Milled pyramids 1.0 mm	5x100	5
	3	Milled pyramids 1.5 mm	5x100	5
	3	Milled pyramids 2.0 mm	5x100	5
	3	Milled circular pattern	5x100	5
	3	Sanded	5x100	5
	3	Sandblasted	5x100	5
	3	Coated with EpoxyTape 0.1 mm	5x100	5
	3	Coated with EpoxyTape 1.0 mm	5x100	5
	3	Coated with grip tape	5x100	5
2	5	Milled pyramids 1.0 mm (offset)	5x100	5
	5	Milled pyramids 1.5 mm (offset)	5x100	5
3	5	Milled pyramids 0.5 mm	6x180	5
	5	Milled circular pattern	6x180	5
4	5	Untreated	6x100	15
	5	Embossed	6x100	15
	5	Milled pyramids 0.5 mm	6x100	15
5	3	Embossed	6x200	15
6	8	Milled pyramids 0.5 mm	6x200	10
	7	Milled pyramids 1.0 mm	6x200	12
7	5	Milled pyramids 1.0 mm	8x300	20

facet size was set to 31 P and the grid space to 9 P. The test procedure and the evaluation were carried out following EN 26891. Both the maximum load $F_{V,test}$ and the stiffness k_s per connector were determined. The stiffness was determined in the range between 10% and 40% of the maximum load in the linear-elastic range. The determined stiffness values are for one connection consisting of two connector plates and thus including the deformation and stiffness of both shear planes (i.e. connector to middle part and connector to side part). The two connector plates of one connection were not connected to each other and the load was transferred via contact pressure. For the tests with the prototype connectors (Series 6 and 7), the connector plates were connected with metric screws but still the load was transferred via contact pressure.

Figure 5.7: Test setup for push-out tests with inclined screws.

5.3.3 Results and discussion

Series 1: 5 screws 5x100 mm

For the first series, small connector plates with only five screws each were chosen to investigate many different surface modifications and their behaviour with inclined screws. The aim of this first test series was to identify suitable surface modifications for the increase of the load-carrying capacity and stiffness.

The failure modes encountered were either tensile failure of one or more screws in the shear plane or pull-out of the screws from the softwood. The tests with connectors with smooth and sanded surfaces and coated with the 1.0 mm EpoxyTape and quartz sand all failed due to the screws reaching their tensile capacity. In contrast, the tests with the connectors with the milled circular pattern and coated with grip tape failed solely due to the screws being pulled out of the softwood. The results indicate a significant increase in the load-carrying capacity of the connection for any surface modification. The only exception here is the tests with grip tape. As with the friction tests, failure occurred in the adhesive bond. Initially, tests were again carried out with a smooth surface to determine comparative values. The mean value of the maximum load of these tests was $F_{V,test}$ = 40.5 kN (per connector) and the stiffness k_s = 16.7$\pm$0.9 kN/mm. For the different pyramid patterns, maximum loads of around 53 kN and stiffness values around 11–16 kN/mm could be determined. The increase in load-carrying capacity is more than 30%, while the stiffness does not increase. Table 5.4 shows the maximum loads and the stiffness values per connector for all surfaces investigated in Series 1.

The load-carrying capacity could be increased because of higher friction in the shear plane and, therefore, an additional load part parallel to the shear plane. This behaviour was especially pronounced with surfaces with protruding features, such as the coated surface or the milled pyramid pattern. This can be explained by the good interlocking of theses surfaces with the softwood. The stiffness of the connections with these protruding surfaces, however, did not increase but rather decrease. For one, the deviation of the test results of the stiffness has to be considered, which was higher than for the ultimate loads. But also, due to the protruding features, the surfaces were further pressed into the softwood during the loading, leading to deformation perpendicular to the shear plane and, thus, lower stiffness values. The high stiffness values on the other hand were determined with surfaces with rather "flat" and even surfaces, such as untreated, sanded, and blasted surfaces. Here, the even surface structures led to a tight fit of the surfaces and no additional deformation perpendicular to the shear plane.

Figure 5.8a shows the load-displacement curves for all surfaces tested in Series 1. Clearly visible is the load increase for all surface modifications. The gradient of the curve and, therefore, the stiffness seems similar for all tests. Figure 5.8b shows a close-up of the local stiffness of each surface in the range of 10–40% of F_{max}. The increase in stiffness is not as pronounced. It becomes visible that surfaces with no protruding features (i.e. sanded and sandblasted, milled 1.0 mm and milled circular) have higher stiffness values than surfaces with protruding features (i.e. milled 1.5–2.0 mm and coated with sand). Like the tests with bolts inserted perpendicular to the shear plane, more significant deformations are needed to impress the protruding surfaces into the softwood thoroughly.

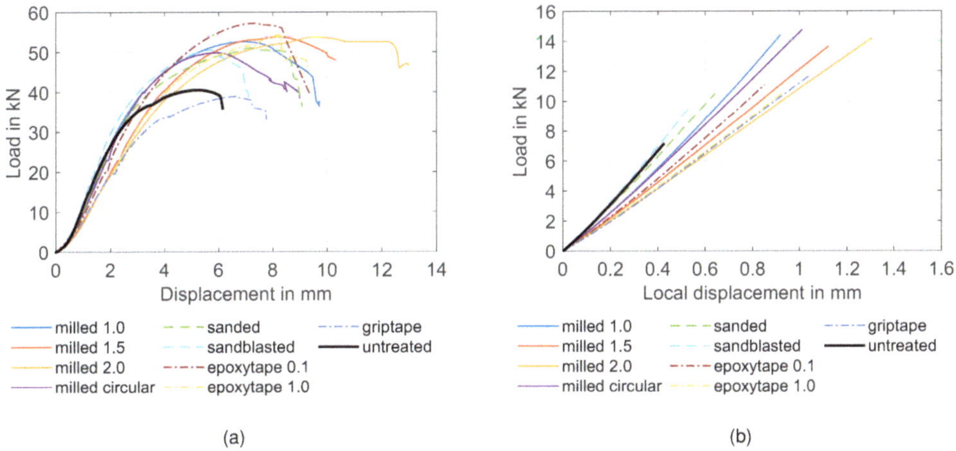

Figure 5.8: Load-displacement curves (a) and local stiffness (b) of tests in Series 1.

Figure 5.9a+b show an uneven spread of the contact pressure and an uneven impression of the modified surface (both milled surfaces: (a) pyramid pattern and (b) circular pattern). Figure 5.9c shows the surface of a test with sand coating. Single sand grains stick to the softwood, and more prominent imprints can be seen. This behaviour is due to single sand grains protruding more than others and, in turn, leading to the connector pivoting over these single grains. The uneven spread of contact force results from a twisting of the connector. This twisting led to an indentation of the lower connector edge into the softwood and a gap in the shear plane at the upper edge, see Figure 5.9d. The indentation of the connector led to additional forces being transferred via direct contact of the bottom edge of the connector plate to the softwood member. To avoid this indentation, the location of the screws in the connector plate was adjusted in Series 2.

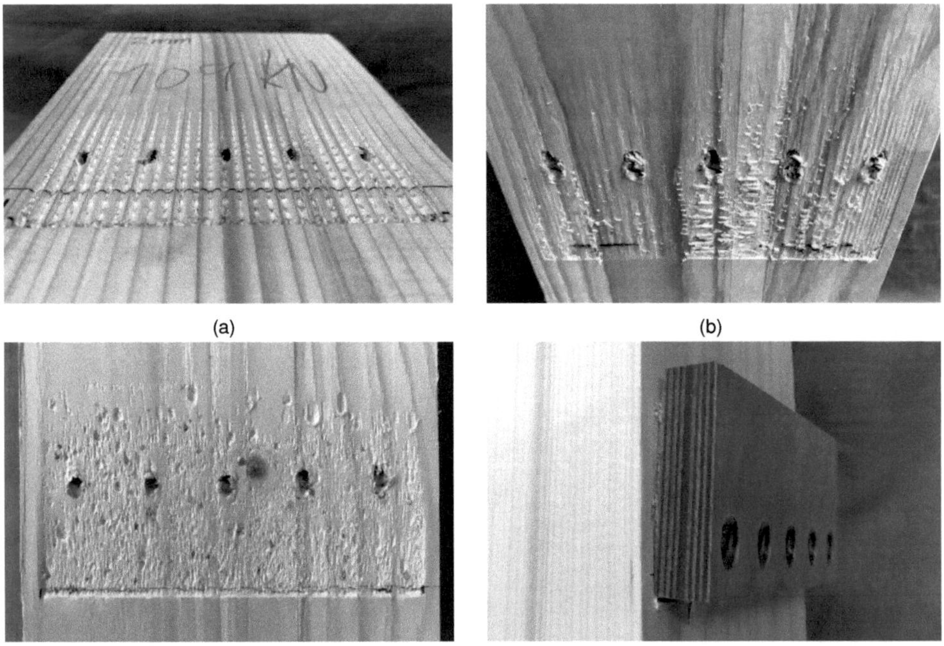

(a) (b) (c) (d)

Figure 5.9: Softwood surface after tests: (a) pyramid pattern; (b) circular pattern; (c) coated with sand; (d) indentation of connector plate.

The results of Series 1 showed that with all investigated surface modifications an increase in load-carrying capacity could be determined. Especially surfaces with protruding features such as the pyramid patterns and the sand-coated surfaces showed significant increases in the ultimate load. The stiffness, however, could not be increased significantly and a large variance of the results was determined. Here, the results showed that rather surfaces with more even features are suitable to increase the stiffness, such as the sanded and sandblasted surfaces. Furthermore, the tests showed an indentation of the connectors and a resulting increase in the ultimate load due to direct contact transfer of the connector plate to the timber member. Based on the results, two pyramid patterns were chosen for the following test series (pyramids with 1.0 and 1.5 mm depth). Also, the location of the screws was adjusted to prevent the connectors from twisting (see Figure 5.10).

Series 2: 5 screws 5x100 mm (offset)

In the second series the connectors were modified by milling a pyramid pattern in the surface (pyramids with 1.0 and 1.5 mm depth). The screw arrangement was varied. An attempt was made to minimise the twisting of the connectors by changing the position of the screws. Therefore, the centre of gravity of the screw group was offset in such a way that the resulting moment from acting force and resisting force was zero (Figure 5.10b+c). This offset reduced the connector's twisting and reduced the average load-carrying capacity. No additional load was transferred via direct contact with the connector's edge (i.e. indentation of the connector and, therefore, direct contact of the connector's edge in the softwood, Figure 5.10a).

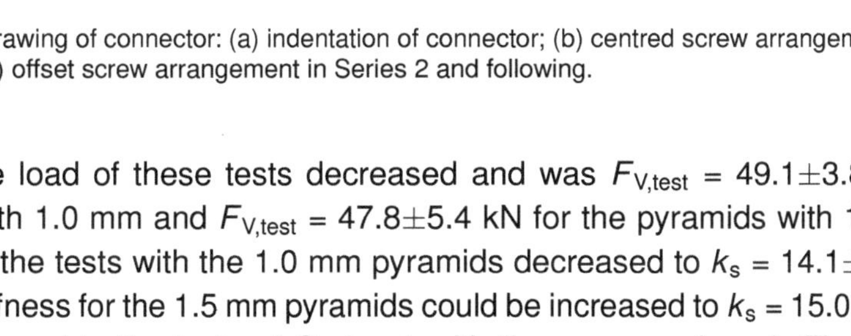

Figure 5.10: Drawing of connector: (a) indentation of connector; (b) centred screw arrangement in Series 1; (c) offset screw arrangement in Series 2 and following.

The ultimate load of these tests decreased and was $F_{V,test}$ = 49.1±3.8 kN for the pyramids with 1.0 mm and $F_{V,test}$ = 47.8±5.4 kN for the pyramids with 1.5 mm. The stiffness for the tests with the 1.0 mm pyramids decreased to k_s = 14.1±2.5 kN/mm, while the stiffness for the 1.5 mm pyramids could be increased to k_s = 15.0±3.2 kN/mm (when compared to the tests of Series 1 with the same surfaces). The decreasing ultimate load can be explained by the prevention of the indentation of the connector's lower edge. Here, no force was transmitted due to direct contact. A performed ANOVA with subsequent Tukey's HSD Test for multiple comparisons, however, showed no significant difference of the ultimate loads ($p = 0.2369$ and $F = 1.64$). The stiffness values for the tests of Series 1 and 2 are similar, which is confirmed by a one-way ANOVA ($p = 0.0701$ and $F = 2.65$).

The failure occurred as before due to reaching the tensile capacity of the screws. The red and blue dashed lines in Figure 5.12 show the load-displacement and local stiffness curves for the two tested surfaces in Series 2 with the offset screw arrangement. For comparison, the curves for the equivalent tests from Series 1 are added with solid lines (same colours for same surfaces). No significant change in the stiffness can be observed. However, the decrease in load-carrying capacity (8–12%) due to the eliminated contact by twisting/indentation of the connector is obvious.

Figure 5.11a shows a connector with offset screw arrangement after the test. The indentation is reduced to a minimum. This reduced indentation is also shown in Figure 5.11b+c with an evenly spread impression of the modified surface into the softwood.

The results of Series 2 showed that with the offset location of the screws, the indentation of the connector could be reduced to a minimum. This, however, also led to an decrease in the ultimate load (8–12%), as no force was transmitted via this contact of the connector edge to the timber member. A uniform pressure distribution in the shear plane was observed, however the stiffness did not further increase. Based on the results, two milled patterns with a more uniform surface structure were chosen for the following test series (pyramids with 0.5 mm depth and the circular pattern) to increase the stiffness. The diameter and length of the screws were enlarged to $d = 6$ mm and $\ell = 180$ mm to ensure tensile failure.

(a) (b) (c)

Figure 5.11: Softwood surface after tests with offset screws: (a) reduced indentation of connector plate; (b) impression of pyramid pattern; (c) impression of circular pattern.

Series 3: 5 screws 6x180 mm

In a third series, the simple connector design from Series 1 and the offset screw arrangement from Series 2 were adopted for tests with longer screws with d = 6 mm and ℓ = 180 mm. With this length, tensile failure of the screw became decisive, and withdrawal of the screws could be neglected. The number of screws was five. Again, only two milled surfaces were investigated, i.e. the milled pyramid pattern with 0.5 mm grooves and the milled circular pattern. For the pyramid pattern, the depth was reduced to 0.5 mm to create a uniform surface with less protruding features to increase the stiffness. This assumption is based on the results of Series 1, which showed that higher stiffness values are possible with even surfaces that will not further impress into the softwood with increasing load in the inclined screws. For the same reasons, the milled circular pattern was chosen. The manufacturing time of the circular pattern is significantly shorter than for the pyramid pattern, but still high friction coefficients were determined in friction tests.

The total load-carrying capacity could be significantly increased to $F_{\text{V,test}} = 84.5 \pm 5.3$ kN for the pyramids with 0.5 mm and to $F_{\text{V,test}} = 80.0 \pm 2.0$ kN for the circular pattern. That is explained by the longer screws being used, thus higher axial loads of the screws. The increase in stiffness, compared to equivalent surfaces from Series 1, was significant with $k_s = 17.8 \pm 1.2$ kN/mm for the tests with the 0.5 mm pyramids, and $k_s = 17.7 \pm 1.8$ kN/mm for the tests with the circular pattern. This can be explained by the surface structure. Both surfaces have a relatively even surface structure that is not further pressed into the softwood during the tests, thus no further deflection perpendicular to the shear plane. Compared to untreated connectors of Series 1, the stiffness increase was only marginal, although screws 6x180 mm were used instead of 5x100 mm. Here, no significant difference can be determined ($p = 0.0678$ and $F = 3.92$).

Figure 5.12 show the load-displacement and local stiffness curves for the two tested surfaces in Series 3 (black lines, solid and dashed), also with an offset screw arrangement. Compared to the other tests with offset screws with ℓ = 100 mm (red and blue lines), the load increase is significant when larger screws are used (d = 6 mm and ℓ = 180 mm). This is due to an increased tensile strength of the screws (+58%, determined in tensile tests of the used screws). An increase in stiffness can be observed with the larger screws. This is due to the surface structure of the tested specimens. The 0.5 mm pyramids and the circular pattern both feature a more even and uniform surface structure that lead to a uniform contact pressure in the shear plane. The surface cannot be further pressed into the softwood when the load on the screws increases. Therefore, higher stiffness values could be determined.

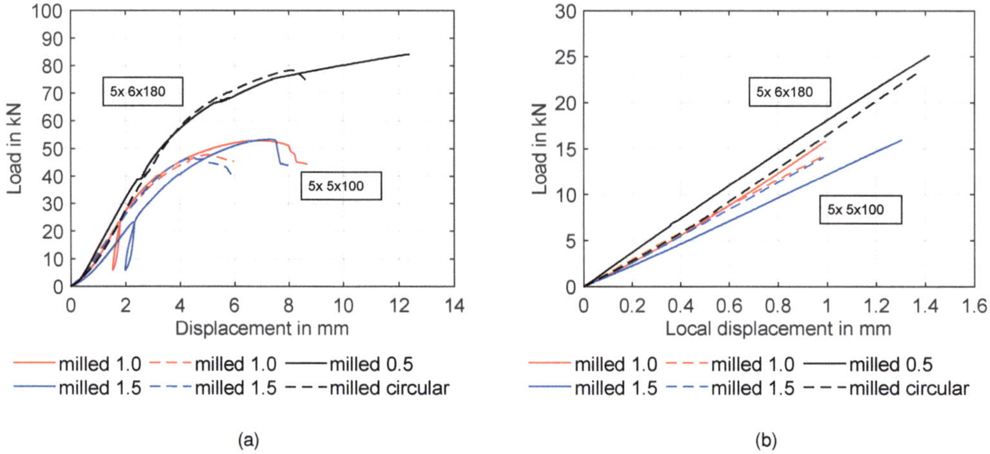

Figure 5.12: Load-displacement curves (a) and local stiffness (b) of tests in Series 2 and 3 (five screws each). Solid lines (red and blue) for tests of Series 1, dashed lines (red and blue) for tests of Series 2 with offset screw arrangement, black lines (solid and dashed) for tests of Series 3 with offset screw arrangement.

The results of Series 3 showed that the ultimate load could be increased with larger screws. These results are due to the increased withdrawal strength of the screws. The stiffness could be increased with even surface structures with uniform features. Compared to the tests of Series 1 with untreated connectors, the load-carrying capacity could be increased due to higher friction by 32% (0.5 mm pyramid pattern) and 25% (circular pattern). Based on the results of all tests with five screws, for the following test series, the number of screws was increased to 15 to investigate group effects. On the one hand, the effective number of screws was to be tested. On the other, block shear failure of a group of screws was investigated.

Series 4: 15 screws 6x100 mm

In order to test for a possible group effect of multiple screws, the number of screws used per connector plate was increased to 15. The screws were arranged in five rows of three screws in the grain direction each and inclined at 45°. Three different surfaces were investigated, i.e. untreated DVW for reference values, an embossed surface with the negative imprint of the pyramid pattern and the pyramid pattern itself with 0.5 mm pyramids. The pyramid pattern was chosen based on the results of the previous tests. The embossed surface was developed after the previous tests, with the aim to further reduce the manufacturing time of the surface modification while simultaneously having a high surface quality throughout the manufacturing process.

The load-displacement curves are shown in Figure 5.15a. All three surfaces show a linear behaviour at the beginning of the tests. However, the curve of the tests with the untreated surface already flattens after reaching 50% of the ultimate load. Whereas the tests with the modified surfaces do not change the gradient before reaching 70–80% of the ultimate load. The ultimate load is reached at $F_{V,test}$ = 119±3 kN for the untreated surface and $F_{V,test}$ = 140±3 kN and $F_{V,test}$ = 153±4 kN for the embossed and milled surface, respectively. The surface modification increases the load-carrying capacity by about 20–30%.

The stiffness analysis shows varying results; see Figure 5.15b. For the untreated and milled surface, the stiffness amounts to k_s = 34±3 kN/mm. Both stiffness values are not significantly different (p = 0.7357 and F = 0.12). The stiffness for the embossed surface could be increased to k_s = 43±9 kN/mm, and differs significantly from the other two surfaces (p = 0.0022 and F = 7.72).

The results for the load and the stiffness can be explained by the structure of the surface: the surface with the pyramid pattern interlocks well with the softwood, increasing the ultimate load. However, the interlocking is due to the surface being pressed into the softwood, thus displacement of the surface. Therefore, the stiffness cannot be increased, as the surface/connector evades the load perpendicular to the loading direction. The embossed surface, however, cannot be further pressed into the softwood, as the surface structure is relatively flat and even. Therefore, the surface/connector does not evade the load, leading to increased connection stiffness.

Figure 5.13 shows the softwood surface after the tests. No group influence of the screws can be detected. The screw arrangement evenly spreads the contact pressure across the shear plane. A clear impression of the embossed surface can be seen, although the surface has no protruding features.

The results of Series 4 confirmed the findings of the previous tests. The load-carrying capacity can be increased due to friction by 29%, when using a surface with protruding features (0.5 mm pyramid pattern). The stiffness can be increased due to friction by 27%, when using an even, uniform surface (embossed pattern). The tests showed no plug shear failure of the group of screws. As the failure mode was withdrawal of the screws, in the following tests, the length of the screws was increased to enforce tensile failure of the screws. Again, these tests were to investigate possible group effects.

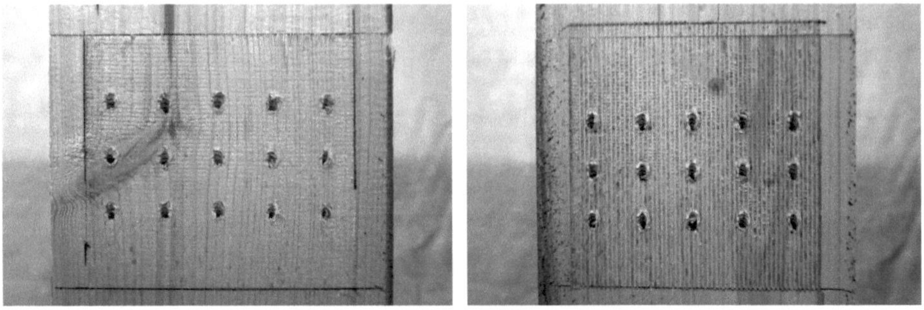

(a) (b)

Figure 5.13: Modified surfaces after tests: (a) embossed inverse pyramid pattern; (b) milled pyramid pattern.

Series 5: 15 screws 6x200 mm

With the same connector plates from Series 4, tests were carried out with longer screws with ℓ = 200 mm. Only three tests were performed, and only the embossed surface was used. The average ultimate load was reached at $F_{V,test}$ = 185$\pm$11 kN. The increase is explained by the use of longer screws, thus increasing the withdrawal capacity. In these tests, failure of the connector plates occurred for the first time (Figure 5.14a+b). In the net cross-section in the top row of screws, compression failure occurred due to buckling of the veneers. Failure of the connector plates was neglected before. Therefore, this failure mode was not considered in the design of the connectors. To avoid compressive failure of the connector plate, the arrangement of the screws can easily be adjusted. A staggered arrangement of the screws would prevent this type of failure and should be considered in the further design of connectors.

Additionally, compressive failure perpendicular to the grain of the softwood side/middle members occurred (Figure 5.14c). This failure mode was coupled with a strong deflection of the connector parts perpendicular to the shear plane. This deflection can be seen in Figure 5.14d. This can be explained by the high amount of screws on a relatively small area and, therefore, also a high contact pressure in this area, reaching the strength of the timber parts. In the design of the connection, compression failure perpendicular to the grain of the timber was not considered. Therefore, the screw arrangement was not adapted to that failure mode. For future connections, the arrangement of the screws should be changed to fully utilise the size of the connector plate.

The load-displacement curve in Figure 5.15a shows similar behaviour to the tests of Series 4. Due to the compressive failure of the connector and the softwood, the curve flattens in the same range as the tests in Series 4 before the ultimate load is reached.

The average stiffness results to k_s = 37.4±3.8 kN/mm and is, therefore, in the same range as the previous tests from Series 4 with the shorter screws, see Figure 5.15b (no significant difference, p = 0.2049 and F = 1.77). This seems reasonable as only the length of the screws was increased, primarily influencing the ultimate load of the connection.

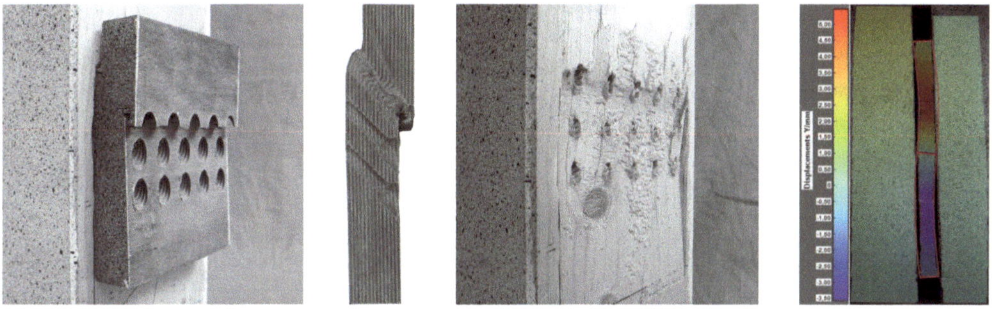

(a) (b) (c) (d)

Figure 5.14: Failure modes: (a+b) compressive failure of DVW connector; (c) compressive failure perpendicular to grain of timber; (d) bending of connector plates.

The results of Series 5 showed an increase in load-carrying capacity due to the longer screws and the higher tensile strength (compared to the withdrawal strength). The stiffness did not further increase, as the same screw's diameter was used. However, the screw's tensile strength could not be fully utilized, and failure of the DVW connector occurred. Based on the preliminary tests of Series 1–5, a first connector prototype was manufactured and tested in the following tests.

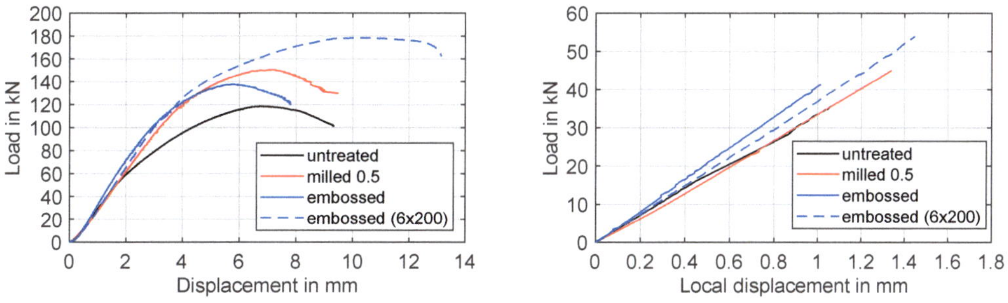

(a) (b)

Figure 5.15: Load-displacement curves (a) and local stiffness (b) of tests in Series 4 and 5.

Table 5.4: Ultimate loads from the tests of Series 1–5 and corresponding stiffness values (mean values).

Series	Surface	$F_{V,test}$ in kN	k_s in kN/mm
1	Untreated	40.5±0.8	16.7±0.9
	Milled pyramids 1.0 mm	52.9±4.0	15.6±0.8
	Milled pyramids 1.5 mm	53.4±1.3	12.2±0.7
	Milled pyramids 2.0 mm	52.8±2.1	11.0±1.0
	Milled circular pattern	49.9±0.2	14.7±1.4
	Sanded	50.3±1.6	16.3±3.0
	Sandblasted	50.6±1.7	17.9±2.5
	Coated with EpoxyTape 0.1 mm	57.8±1.2	13.6±0.7
	Coated with EpoxyTape 1.0 mm	52.3±2.4	11.4±0.7
	Coated with grip tape	39.0±2.1	11.4±0.9
2	Milled pyramids 1.0 mm (offset)	49.1±3.8	14.1±2.5
	Milled pyramids 1.5 mm (offset)	47.8±5.4	15.0±3.2
3	Milled pyramids 0.5 mm	84.5±5.3	17.8±1.2
	Milled circular pattern	80.0±2.0	17.7±1.8
4	Untreated	119±3	33.5±2.7
	Embossed	140±3	42.6±9.0
	Milled pyramids 0.5 mm	153±4	34.0±3.5
5	Embossed	185±11	37.4±3.8

Series 6: 10–12 screws 6x200 mm

After the tests with varying numbers and arrangements of the screws, push-out tests with connector prototypes made of densified veneer wood were performed. The objective of the first prototype was to reach an ultimate load of approximately 180 kN. For the surface modification a pyramid pattern with 0.5 mm grooves was chosen. The number of screws was ten, arranged in five rows of two screws in grain direction. Drawings of the connectors are given in Appendix A.2, Figures A.11–A.13. The push-out tests with the prototype v1 achieved an average load of $F_{V,test}$ = 151±5 kN and a stiffness of k_s = 25.0±4.0 kN/mm. In the tests, six out of eight test specimens failed after reaching the tensile strength of the screws, and two test specimens failed after reaching the compressive strength of the DVW. The load-displacement curves are given in Figure 5.16. Shown with dashed lines are the two tests with connector failure. The ultimate load and stiffness are still similar to the other tests, where the screws failed.

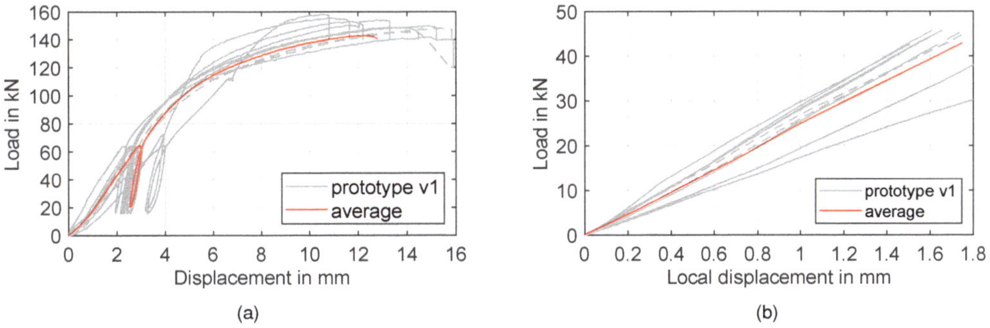

Figure 5.16: Load-displacement curves (a) and local stiffness (b) for tests with prototype v1 and milled pyramid pattern.

Figure 5.17a shows a specimen with failed screws after the test. In Figure 5.17b+c, the even distribution of the contact pressure can be seen. The inclined screws press the modified surface firmly and equally into the softwood, although the screws are not equally spread over the contact area. Additionally, only minimal indentation of the connector plate into the softwood can be seen.

The tests showed that the aim of an ultimate load of 180 kN was not reached. Tensile failure perpendicular to the grain of the DVW connector plate occurred, after reaching the ultimate load and displacement. Therefore, a second prototype was developed with various changes. The number of screws was increased to twelve, and the surface modification was changed to a pyramid pattern with 1.0 mm grooves. The position of the screws was adapted to four rows of three screws in grain direction. Additionally, mounting screws perpendicular to the shear plane were arranged to fix the connector plate before installing the inclined screws.

(a) (b) (c)

Figure 5.17: Specimens with prototype v1 after tests: (a) tensile screw failure at right top plate; (b) impression of bottom plate; (c) impression of top plate.

In the subsequent tests with the second prototype v2, an average load-carrying capacity of $F_{V,test}$ = 173±8 kN and a stiffness of k_s = 33.3±2.4 kN/mm could be determined. The main difference between connector v2 and connector v1 was that 12 screws were used instead of 10, and they were arranged differently. Due to large indentations of the prototype connectors v2 in the softwood side members, two further tests were carried out with 5-ply cross-laminated timber (CLT). Here, the average maximum load was $F_{V,test}$ = 186 kN and especially the stiffness of the connection could be significantly increased by the transverse layers with k_s = 63.5 kN/mm. The failure was due to reaching the tensile capacity of the screws in all tests except the last test with CLT, where rolling shear failure/tensile failure perpendicular to the grain occurred.

Figure 5.18a shows the averaged force-displacement diagrams. The red line shows the behaviour of the tests that failed due to failure perpendicular to the grain. The curve flattens already at an early stage at about 40% of the ultimate load. The blue line shows the behaviour of the tests that failed suddenly due to the tensile failure of the screws. The behaviour of the connection is stiffer than the test with timber failure, but the ultimate load is only slightly higher. The black line shows the behaviour of the tests with CLT members instead of GLT. The ultimate load could be further increased, and the stiffness could be increased significantly. Figure 5.18b shows the stiffness of the two prototype connectors. It can be seen, that the stiffness increases when more screws are used (blue line compared to red line) and when the compressive strength perpendicular to the grain of the timber increases (black line compared to blue line).

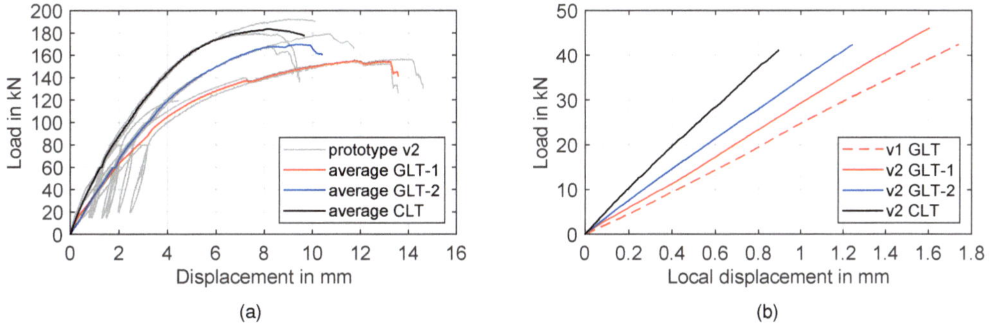

Figure 5.18: Load-displacement curves (a) and local stiffness (b) for tests with prototype v2 and milled pyramid pattern.

Figure 5.19a shows the effectiveness of the inclined screws, which press the surface evenly into the softwood right up to the edge of the connector plate. A slight rotation of the fastener can be seen. This rotation becomes more visible in Figure 5.19b, with the top edge of the fastener pressing into the softwood. The crushing of the wood fibres occurs when the inclined screws suddenly fail, and the fastener is abruptly pressed into the wood. Figure 5.19c shows again the even distribution of the contact pressure (here for the tests with CLT) and the more evenly arranged screws.

(a) (b) (c)

Figure 5.19: Specimens with prototype v2 after tests: (a) impression of pyramid pattern; (b) crushing of fibres at top plate; (c) impression of pyramid pattern in CLT.

During the tests, large displacements of the connectors were observed. This displacement led to plastic deformation of the assembly screws, which were inserted perpendicular to the shear plane (Figure 5.20a). After opening the test specimens, a deviation of the screws from their supposed position became evident (Figure 5.20b).

However, the screws did not touch or damage each other (unlike for the tests with prototype v3, see Figure 5.20c).

(a) (b) (c)

Figure 5.20: Fully-threaded screws after the tests: (a) plastic hinges in the mounting screws; (b) deviation of inclined screws; (c) damaged screws.

Table 5.5: Results for load $F_{V,test}$ and stiffness k_S in insertion direction of the prototype connectors.

Series	Prototype	Timber	Surface	$F_{V,test}$ in kN	k_S in kN/mm
6	v1	GLT	Milled pyramids 0.5 mm	151±5	25.0±4.0
	v2	GLT[1)]	Milled pyramids 1.0 mm	156±2	30.3±3.3
	v2	GLT	Milled pyramids 1.0 mm	173±9	33.3±2.4
	v2	CLT	Milled pyramids 1.0 mm	186±9	63.5±8.7
7	v3	GLT[1)]	Milled pyramids 1.0 mm	443	82.0
	v3	GLT[2)]	Milled pyramids 1.0 mm	496±4	80.9±4.3
	v3	CLT	Milled pyramids 1.0 mm	503	105

1) compressive failure perpendicular to the grain of the softwood members

2) softwood members reinforced with fully threaded screws

The results of Series 6 showed connectors made of DVW are feasible and an alternative to conventional system connectors. Based on the results, the aim in the next series was to develop a connector for loads up to 500 kN.

Series 7: 20 screws 8x300 mm

The objective of this test series was to develop and test a prototype connector made of DVW for loads up to 500 kN. Therefore, the number of screws was increased to 20. The screws were enlarged to d = 8 mm and ℓ = 300 mm. The dimensions of the connector were also enlarged accordingly. In the first test, the estimated load could not be reached because the two side members beneath the connector plates failed due to compression perpendicular to the grain. In the subsequent tests, the softwood members were reinforced with fully threaded screws 8x160 mm. The screws were countersunk a few millimetres so that the milled pyramid pattern could be pressed freely into the softwood. As a result, the DVW connector plates failed due to compression parallel to the grain in two tests. Only in one test, the tensile capacity of the screws was reached. The average ultimate load for the test where timber failure occurred was $F_{V,test}$ = 443 kN and the stiffness k_s = 82.0 kN/mm. The average ultimate load for the reinforced tests was $F_{V,test}$ = 496$\pm$4 kN per connector and shear plane and the corresponding stiffness k_s = 80.9$\pm$4.3 kN/mm. In order to do without the reinforcement screws, a final test specimen was made of 5-ply CLT. In the test, only a slightly higher ultimate load could be achieved with $F_{V,test}$ = 503 kN. However, the connection stiffness could be increased by almost 30% to k_s = 105 kN/mm. Figure 5.21 shows the load-displacement diagram. In red is the test without reinforcement that failed due to compressive failure perpendicular to the softwood grain. Visible is the plastic behaviour with large deformations at a constant load. In blue are the tests with additional reinforcement screws. The initial behaviour of the curve is very identical, which is also shown in the similar stiffness values. Due to the reinforcement, the ultimate load can be slightly increased. In black is the test with side and middle members made of CLT. The ultimate load is identical (same failure mode). However, the gradient of the curve is significantly steeper, and therefore the stiffness is higher.

The firm imprint of the pyramid pattern into the softwood can be seen in Figure 5.22a. The modified surface pressed evenly in all layers with the CLT parts, regardless of their grain direction. In addition, large displacements of the connectors were observed in these tests. After the opening of the specimens, the bending of the inclined screws was visible as in Figure 5.22b. This bending may explain the discrepancy between the estimated and maximum loads. When disassembling the connectors after the test, damaged screws were discovered; see Figure 5.20c. When inserting the inclined screws, there must have been a collision with a mounting screw inserted perpendicular to the shear plane, so that the thread of both screws was damaged. The mounting screw that was inserted first was hit at one point. The inclined screw that was inserted later had a damaged thread from the tip to the point where the two screws touched. This problem was already pointed out by Frese & Jordan [82], and an appropriate taper model for the minimum distance was proposed. However, this was not applied here in the choice of minimum distances, as the approximately 60 mm long guidance of the screws through the connector plate was considered sufficient to avoid contact.

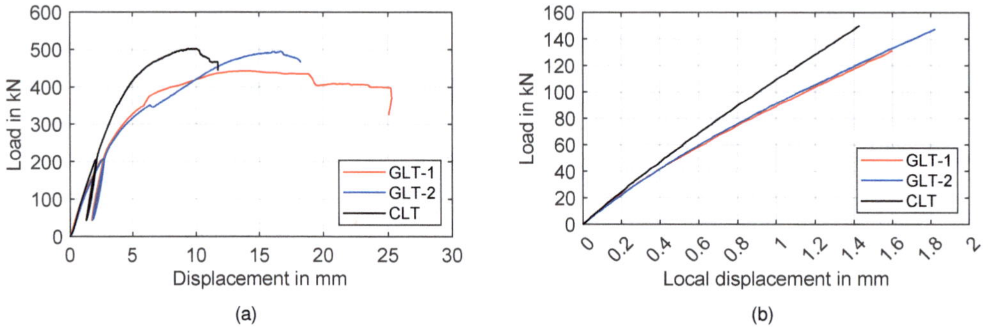

Figure 5.21: Load-displacement curves (a) and local stiffness (b) for tests with prototype v3 and milled pyramid pattern.

Figure 5.22: Specimens with prototype v3 after tests: (a) impression of pyramid pattern in CLT; (b) bending of screws; (c) screws for compressive reinforcement perpendicular to grain.

5.3.4 Main beam to secondary beam tests

Test programme and setup

Based on the push-out tests, large-scale main beam to secondary beam tests were carried out to investigate the behaviour of connectors with modified surfaces and increased friction in the shear plane on the overall system. Five tests were conducted with the prototype connector v2 and three with the heavy-duty prototype v3. The test setup comprised a secondary beam connected at both ends to a main beam via the DVW connectors. Due to material requirements, it was decided to use a three-point bending test rather than a significantly more material-intensive four-point bending test. The test setup can be seen in Figure 5.23a. Glulam GL 24h was used for the main and secondary beams. The length of the secondary beams was chosen so that load propagation at 45° starting from the load application would not interfere with the inclined screws of the connectors. Dimensions and locations of the connectors can be seen in the drawings in Figure 5.23b+c.

To prevent tensile failure perpendicular to the grain of the secondary beam, the connector plates were fixed to the lower edge of the secondary beam. On the main beam, the connectors were arranged to maintain the ratio $h_e/h \geq 0.7$, and thus no analytical check for tensile failure perpendicular to the grain was necessary. In addition, the components were reinforced accordingly. The measures included reinforcement for compression perpendicular to the grain beneath the load application and at the support, as well as a reinforcement for tension perpendicular to the grain just behind the tips of the inclined screws of the connectors (Figure 5.27c). Additionally, the main beams were reinforced with fully threaded screws under the connector plates in order to exclude premature compression failure perpendicular to the grain. As before in the push-out tests, the fully threaded screws were countersunk by 2.5 mm so that the milled pyramid pattern could still be pressed into the softwood.

The test procedure was the same as for the push-out tests with the unloading loop, according to EN 26891. During the tests, the vertical relative displacement of the secondary beam to the main beam at the height of the connectors was measured with inductive displacement transducers, as well as the horizontal displacement of the secondary beam at its upper and lower edge (measuring rate of 5 Hz). The horizontal relative displacement was used to calculate the rotation of the secondary beam. The main beams were secured against rotation with fork bearings.

Results and discussion

Connector prototype v2 with milled pyramid pattern

After the tests, the maximum load per connector, independent of deformation, and the stiffness between 10% and 40% of the maximum load were evaluated. The results are

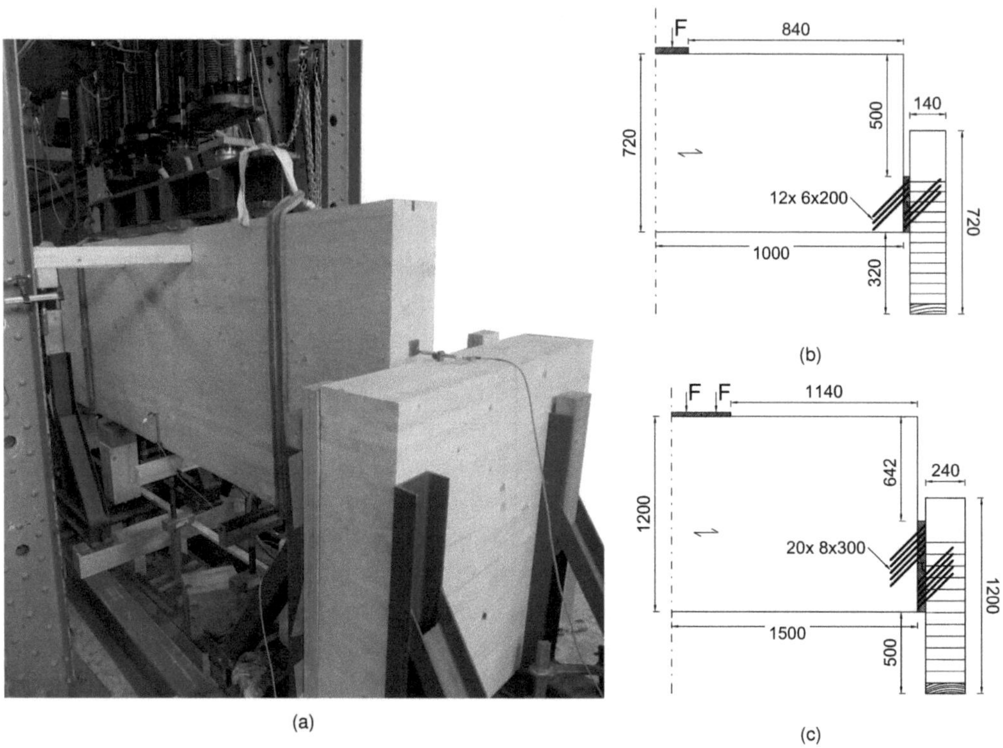

Figure 5.23: Test setup for main beam to secondary beam tests (a) and drawings of test setup for tests with connector v2 (b) and connector v3 (c).

given in table 5.6. On average, the tests reached a maximum load of $F_{V,test}$ = 172±7 kN, roughly corresponding to the test results of the push-out tests. Statistically there is no significant difference between the two test series (p = 0.9928 and F = 0.00). The load-displacement diagrams are given in Figure 5.24. As can be seen, the behaviour is very similar to the push-out tests. The mean stiffness was also close to the results of the previous push-out tests with k_s = 31.1±3.2 kN/mm. Again, statistically there is no significant difference (p = 0.8765 and F = 0.02). In the first test, an error occurred in the data acquisition, so the test was stopped at a load of approximately 90 kN. The same test specimen was tested again in the second test, which explained the slightly higher stiffnesses. The rotation of the secondary beam at its ends was, on average, only φ_{test} = 0.38 degrees. This low rotation is because of the short insertion nozzle. The upper connector plate was free to rotate and thus did not undergo the severe twisting of the lower connector plate, as seen in commercial system connectors.

Figure 5.25a shows the excellent impression of the modified surface into the softwood. The reinforcement screws did not obstruct the pyramids to interlock with the timber and to build up the higher friction. A vigorous twisting of the lower connector plates occurred because the connector on the main beam pressed very firmly into the timber despite the compression reinforcement (see Figure 5.25b). This twisting caused the failure of three of the four tests. Only in the last test did failure of the screws occur, which were pulled out of the end grain of the secondary beam; see Figure 5.25c. Thus, similar capacities and stiffness values could be determined as in the push-out tests, but the failure modes differed significantly. This difference can probably be attributed to the fact that due to the twisting of the secondary beam, the force is no longer transmitted centrally in the connector. An eccentricity was created, which caused the connector to twist and be pressed into the face grain of the main beam. This indentation was as before observed in Series 1 of the push-out tests with five screws.

The same ultimate load and stiffness was expected, as the same connectors were used. The slightly lower gradient of the load-displacement curve after the unloading loop can be explained by the twisting of the connector plate and, thus, indenting of the bottom connector plate in the main beam. This also explains the higher displacement when reaching the ultimate test load.

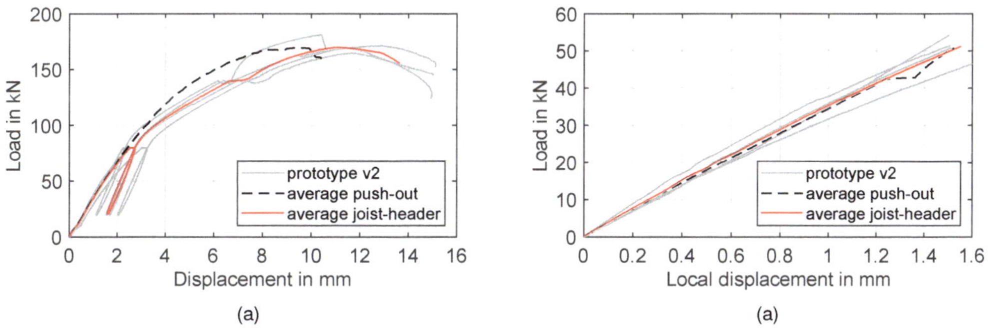

Figure 5.24: Load-displacement curves (a) and local stiffness (b) for main beam to secondary beam tests with connector v2 and comparison to push-out tests.

(a) (b) (c)

Figure 5.25: Impression of modified surface (a), twisting of the connector (b) and withdrawal of screws (c).

Table 5.6: Results for load $F_{V,test}$ and stiffness k_s for main beam to secondary beam tests.

Series	Prototype	Surface	$F_{V,test}$ in kN	k_s in kN/mm
6-JH	v2	Milled pyramids 1.0 mm	172±7	31.1±3.2
7-JH	v3	Milled pyramids 1.0 mm	394±33	71.6±2.6

Connector prototype v3 with milled pyramid pattern

In the tests with the larger prototype v3, it was impossible to exploit the connectors' full potential and reproduce the results of the push-out tests. On average, the tests resulted in a maximum load of $F_{V,test}$ = 394±33 kN and a stiffness of k_s = 71.6±2.6 kN/mm. A one-way ANOVA showed a significant difference for the load ($p = 0.0063$ and $F = 27.48$) and for the stiffness ($p = 0.0002$ and $F = 29.34$). Figure 5.26a shows the averaged load-displacement diagram of the main beam to secondary beam tests and, in comparison, the averaged load-displacement diagram of the push-out tests. Figure 5.26b shows the according stiffness. The behaviour of the curves is similar up to a displacement of around 8 mm. Here, the curve of the main beam to secondary beam tests flattens before reaching the maximum. This difference can be explained by different failure modes that occurred: failure of the screws for the push-out tests in contrast to failure of the timber perpendicular to the grain for the main beam to secondary beam tests. Interestingly, the load and stiffness of the push-out test without reinforcement that failed perpendicular to the grain is still higher than for the main beam to secondary beam tests.

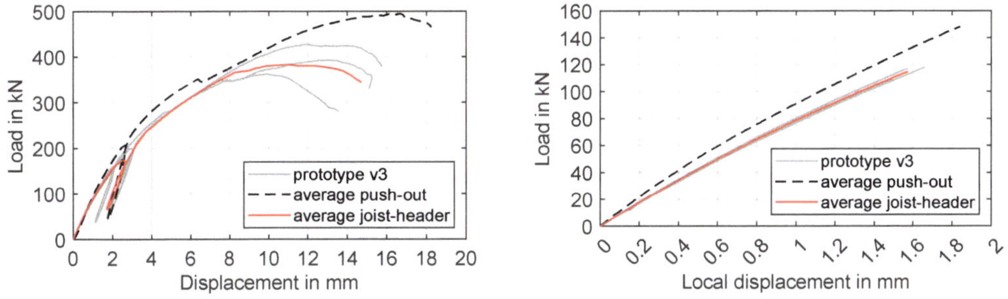

(a) (b)

Figure 5.26: Load-displacement (a) and local stiffness (b) for main beam to secondary beam tests and comparison to push-out tests.

As with the tests before, the inclined screws spread the contact pressure evenly, and an even impression of the modified surface can be seen in Figure 5.27a. The reinforcement screws perpendicular to the grain do not obstruct the pyramid pattern from being pressed into the softwood. In contrast to the push-out tests, tensile failure perpendicular to the grain occurred at the end of the secondary beam, namely in the area of the inclined screws (Figure 5.27b). In the third and last test, additional fully threaded screws were screwed in between the inclined screws of the connectors as tensile reinforcement (Figure 5.27c). This reinforcement again increased the load-carrying capacity slightly, but the connection fell short of expectations.

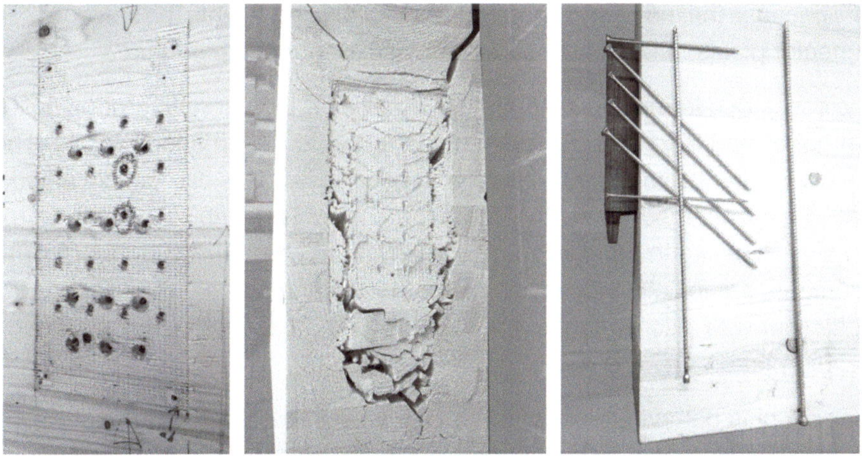

(a) (b) (c)

Figure 5.27: Impression of pyramid pattern (a), tensile failure perpendicular to grain and rolling shear failure of the secondary beam (b) and reinforcement against tension perpendicular to grain (c).

5.4 Summary and conclusions

Tests with fasteners inserted perpendicular to the shear plane and inclined to the shear plane were performed. In all tests, the surface of the DVW or steel was modified. Eleven tests with bolts inserted perpendicular to the shear plane were performed in total. The tests were double-shear steel-to-timber tests. The steel parts were modified with a milled surface and a coated surface. When using a modified steel surface, the results showed only a slight increase in load-carrying capacity (+16–24%) but no increase in stiffness.

A total of 90 tests were carried out with inclined screws and modified surfaces to increase the friction in the shear plane. The tests were single-shear DVW-to-timber tests. The DVW surface was modified in many ways, mainly by milling. The results showed increased load-carrying capacities for all modified surfaces. A significant increase in the load-carrying capacity of 43% was determined with a coated surface. Due to their immediate loading in the axial direction, the inclined screws were able to press the modified surfaces into the softwood and activate higher friction. However, the tests showed that the determined increases of the coefficient of friction do not apply to connections to the same extent.

The increase in stiffness was lower with only +10%, and determined with different surfaces than the increased load capacity. However, in one test series an increase of +27% was reached with an embossed surface. The increase in stiffness was only marginal because the modified surfaces were further pressed into the softwood, causing the connector plates to be displaced perpendicular to the shear plane.

It can be concluded from the tests that "flat" surfaces without (great) protruding features perform better and lead to a higher increase in the load-carrying capacity. This conclusion is congruent with the results of the friction tests. This can be explained by the fact that with the "flat" surfaces, the contact pressure is distributed evenly across the shear plane, resulting in more even friction. With the other surfaces, more significant displacements are initially required to press the surface completely into the softwood. This behaviour was also evident in the tests with the fasteners inserted perpendicular to the shear plane.

Furthermore, this behaviour leads to the conclusion that a surface modification is of minor importance for the overall load-carrying capacity of the connection in the case of connections with fasteners inserted perpendicular to the shear plane. Here, the initial torque on the fasteners is of superior importance, resulting in high initial stiffness.

Overall it can be concluded that the load-carrying capacity can be increased with surface modification but for the stiffness it all depends on the type of the surface modification. However, in the case of beam-to-beam connectors the stiffness might also be of lower importance as these structural systems usually are statically determinate.

6 Modelling and comparison to experimental results

6.1 Introduction and objectives

To incorporate the benefits of increased coefficients of friction in connections into the design of such connections, an analytical calculation model is presented and analysed for its predictive power. In doing so, modelling techniques already shown in the literature are used and checked to see whether they are also valid for high friction coefficients. Failure types that have occurred in tests that have not been considered before will also be implemented in the design equations.
The analytical model aims to provide the designer of connections with increased friction in the shear plane with a straightforward approach to calculating the load-carrying capacity of such connections.

To determine the deformation of the connection (which is not covered by the analytical model), a numerical two-dimensional model was created. This two-dimensional model is based on springs connecting the parts. The springs represent the various screws in the connection. The springs' input parameters are based and validated on independent small-scale tests to ensure transferability to other connections not tested here.
This 2D model aims to provide the designer with a tool to accurately predict the deformation of the connection.

To identify the stresses in the connector itself (which is not covered by the 2D model), a three-dimensional model was developed. This three-dimensional model is based on volume elements and interaction properties to reproduce the load-carrying and deformation behaviour as accurately as possible. This further increases the complexity of the model. Due to the small amount of available test data, the input parameters are based and validated on independent small-scale tests to ensure transferability.
The 3D model aims to supplement the above models with aspects that these models cannot depict (mainly the deformation of the connection). In addition, this model should offer the possibility of optimising the shape of the connectors with modified surfaces.

The modelling approaches for all three models follow publications from the literature to ensure that these techniques have already been tested and to compensate for the lack of additional test data for validation.

6.2 Analytical model

6.2.1 Design equations

6.2.1.1 Load-carrying capacity of axially loaded screws

The design of connections with screws mainly considers three types of failure. That is 1) failure of the bond between the screw's thread and the wood when reaching the withdrawal capacity, leading to the screw being pulled out of the wood. 2) Failure of the screw itself when reaching the tensile capacity, leading to screw rupture. 3) Failure of the wood adjacent to the screw's head when reaching the head pull-through capacity, leading to the screw being pulled through the wood on the side of the screw head. In addition, the possibility of 4) buckling failure of the screw should be investigated in the case of compressive loading. For a group of screws combined with steel parts, there is also the possibility of 5) shear failure along the circumference (block shear failure).

The failure cases 3) to 5) are not considered further in the analytical model. As system connectors are made of either steel or aluminium or, in this case, densified veneer wood, the head pull-through capacity will not become decisive, as the screws will rupture before (see also results of pull-through tests in [3]). Buckling is not considered because the screws are loaded in tension. Block shear failure is checked using the equations in [83]. The determined block shear capacities range from 100 kN for the smallest to 1150 kN for the largest connectors. Therefore, block shear is not further considered because the penetration depth of the screws is large enough.

The load-carrying capacity of connections with dowel-type fasteners is usually calculated according to the theory of Johansen [15]. However, only the load perpendicular to the axis of the fastener is considered. In Figure 6.1, a connection with inclined arranged screws can be seen. The screws are inclined by the angle ε to the plane of the connector or shear plane. For reasons of equilibrium, the horizontal component $\sin\varepsilon \cdot F_{ax}$ of the screw's force creates a compressive force perpendicular to the shear plane. This compressive force, in turn, activates friction in the shear plane. This effect is also called the *rope effect*.

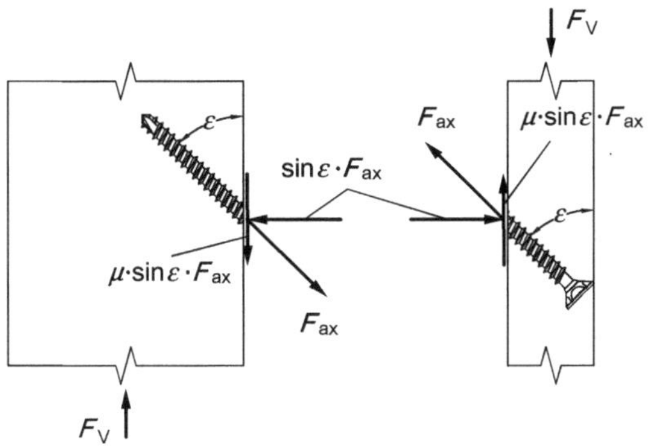

Figure 6.1: Force diagram with acting forces of a connection with inclined screws.

Bejtka & Blaß [7] showed that additional forces could be transmitted via friction between the components already with a slight inclination of the screws. This additional force is considered in the design of laterally loaded fasteners (ε = 90°) with the factor $\Delta F_{V,Rk}$. As the angle ε decreases, the share of axial load in the total load-carrying capacity increases while the share of lateral load decreases. Due to different stiffness values in axial and lateral loading – the difference is about one order of magnitude – at an angle ε = 45°, practically the entire load is transferred by axial loading of the screw, and the share from lateral loading can be neglected [84].

The total load-carrying capacity F_V of the connection is thus calculated analogously to Figure 6.1 as the sum of the part of the force in the screw parallel to the shear plane and the friction force in the shear plane. Equation 6.1 follows research conducted by Blaß et al. [85] and more recently by Krenn [86].

$$F_V = n_{ef} \cdot F_{ax} \cdot (\cos\varepsilon + \mu \cdot \sin\varepsilon) \tag{6.1}$$

with	n_{ef}	effective number of screws
	F_{ax}	axial capacity of one screw
	ε	inclination angle of the screw with respect to the shear plane
	μ	coefficient of friction

Withdrawal capacity

Two different formulae are considered for the withdrawal capacity of screws. Equation 6.2 by Blaß et al. [85] determines the withdrawal capacity as a function of the inclination angle ε. Further input parameters are the outer diameter *d*, the penetration length ℓ_{ef} and the density ρ. The equation is based on over 700 tests and can be applied to self-drilling wood screws from other manufacturers if the geometric properties are similar to those of the screws studied. It can be seen that the equation was determined deterministically, and therefore, the parameters must be entered with their specific units.

$$F_{\mathrm{ax}} = \frac{0,6 \cdot \sqrt{d} \cdot l_{\mathrm{ef}}^{0,9} \cdot \rho^{0,8}}{1,2 \cdot \cos^2 \varepsilon + \sin^2 \varepsilon} \tag{6.2}$$

with F_{ax} withdrawal capacity of one screw [N]

d diameter of the screw [mm]

ℓ_{ef} effective penetration length of the screw [mm]

ρ density [kg/m³]

ε inclination angle of the screw with respect to the shear plane [°]

Equation 6.3 of Frese et al. [87] also calculates the withdrawal resistance of self-drilling wood screws and was determined through regression calculations based on 1850 withdrawal tests. The same input parameters apply (outer diameter *d*, penetration length ℓ_{ef}, density ρ), except for the inclination angle ε. In addition, great care must be given to the proper units. If the geometric properties of the screws are similar, Equation 6.3 estimates the withdrawal resistance more accurately than previously calculated but is only valid for penetration depths up to 140 mm.

$$\ln F_{\mathrm{ax}} = 6.739 + 0.03257 \cdot l_{\mathrm{ef}} + 2.148 \cdot 10^{-4} \cdot d \cdot \rho - 1.171 \cdot 10^{-4} \cdot l_{\mathrm{ef}}^2 \tag{6.3}$$

with F_{ax} withdrawal capacity of one screw [N]

ℓ_{ef} effective penetration length of the screw [mm]

d diameter of the screw [mm]

ρ density [kg/m³]

Other models for calculating the withdrawal capacity were not considered, as the design of the connection with inclined screws and increased friction is intended to achieve tensile failure of the screws and, thus, full utilisation of the screws. It is also important to note here that both equations were determined in tests with screws loaded in their axial direction. However, the equations are used in the herein proposed design model to predict the withdrawal capacity for inclined screws loaded perpendicular to their axial direction (which results in an axial load of the screws due to the inclination). It was not further investigated if the withdrawal capacity of inclined screws (axial load and perpendicular load in shear) differs from the withdrawal capacity of solely axially loaded screws. Again, it is assumed that full utilisation of the screws, thus tensile failure, is the preferred failure mode of the connection.

Tensile capacity

The tensile capacity of screws is dependent on the geometry and material of the screws and can, therefore, easily be determined experimentally. Tensile tests of the respective screws were performed, and the results are given in Table 6.1.

Table 6.1: Tensile capacity of screws in kN (determined experimentally).

Screw	5x100	6x100	6x200	8x300
Number of tests	12	12	15	5
MEAN	8,96	15,1	14,2	24,1
SD	0,06	0,43	0,10	0,14
COV	0,6%	2,9%	0,7%	0,6%

6.2.1.2 Load-carrying capacity of connector

The total load-carrying capacity of the connector itself also has to be considered. This failure case is rare for standard system connectors made of steel or aluminium. However, for the connectors made of densified veneer wood (DVW), the material's compressive strength is much lower than for conventional materials. The load-carrying capacity can be determined in tests. If the compressive strength $f_{c,0}$ of the used material is known, the load-carrying capacity can be calculated using the smallest net area A_{net} (see Figure 6.2) in the loading direction, see Equation 6.4. This approach is an assumption and not further validated in tests.

$$F_V = A_{net,DVW} \cdot f_{c,0,DVW} \tag{6.4}$$

with $A_{net,DVW}$ net cross-sectional area of the connector
$f_{c,0,DVW}$ compressive strength parallel to the grain of the connector

6.2.1.3 Load-carrying capacity of timber

Finally, the load-carrying capacity of timber perpendicular to the grain must be considered. The layout of the screws for all connector prototypes provides for many screws in a small area. This layout leads to significant stresses perpendicular to the grain locally under the connector plates. The load-carrying capacity can be calculated with Equation 6.5, taking into account the loaded area of the header $A_{c,90,H}$ (including 30 mm at each side in fibre direction, see Figure 6.2), the header's compressive strength perpendicular to the grain $f_{c,90,H}$, and the parameter $k_{c,90,H}$, according to EC 5.

$$F_V = A_{c,90,H} \cdot f_{c,90,H} \cdot k_{c,90,H} \cdot \left(\mu + \frac{1}{\tan \varepsilon}\right) \tag{6.5}$$

with $A_{c,90,H}$ area under the connector plate at the header
$f_{c,90,H}$ compressive strength perpendicular to the grain of the header
$k_{c,90,H}$ coefficient for compression perpendicular to the grain

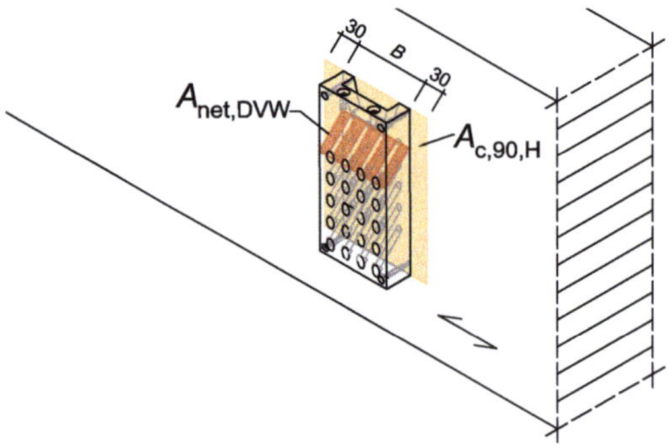

Figure 6.2: Loaded net area A_{net} in connector plate and loaded area of the header $A_{c,90,H}$ under the connector plate.

6.2.1.4 Total load-carrying capacity of the connection

The total load-carrying capacity of the connection with inclined screws results from the minimum of Equation 6.6. For the calculation of $F_{ax,R}$, the density was determined after the tests on each side and middle member and was in the range of 416–489 kg/m^3. The moisture content was in the range of 9.6–13%. The actual density and moisture content for each specimen is documented in Appendix A.2. The static coefficients of friction μ_{stat}, determined for each surface in Chapter 4, were taken for the friction coefficients. The number of screws was calculated as n_{ef} = 0.9·n according to ETA-19/0553 [118]. Irrespective of this, this is also recommended based on tests in [86].

$$F_V = min \begin{cases} n_{ef} \cdot F_{ax,R} \cdot (\mu \cdot \sin\varepsilon + \cos\varepsilon) \\ n_{ef} \cdot F_{tens,R} \cdot (\mu \cdot \sin\varepsilon + \cos\varepsilon) \\ A_{net,DVW} \cdot f_{c,0,DVW} \\ A_{c,90,H} \cdot f_{c,90,H} \cdot k_{c,90,H} \cdot \left(\mu + \frac{1}{\tan\varepsilon}\right) \end{cases} \tag{6.6}$$

with	n_{ef}	effective number of fasteners
	$F_{ax,R}$	withdrawal capacity of a screw
	$F_{tens,R}$	tensile capacity of a screw
	μ	coefficient of friction
	ε	inclination angle of the screw with respect to the shear plane
	$A_{net,DVW}$	net cross-sectional area of the connector
	$f_{c,0,DVW}$	compressive strength parallel to the grain of the connector
	$A_{c,90,H}$	area under the connector plate at the header
	$f_{c,90,H}$	compressive strength perpendicular to the grain of the header
	$k_{c,90,H}$	coefficient for compression perpendicular to the grain

6.2.2 Comparison with experimental results

The test results of Series 1 and 2 with screws 5x100 mm are given in Table 6.2. Additionally, the analytical load-carrying capacities when reaching the withdrawal strength of the screws, which was decisive for the calculation, are shown. The expected values, $F_{V,exp}$, were calculated with Equation 6.6. Hereby, the withdrawal strength $F_{ax,R}$ was calculated with Equation 6.2 or Equation 6.3. For comparison, the ratio of the expected value, $F_{V,exp}$, to the ultimate load from the tests, $F_{V,test}$, was calculated. The comparison provided a smallest ratio value of 0.91 and a mean ratio value of 1.14 when using Equation 6.2 for F_{ax} and a smallest ratio value of 0.87 and a mean ratio value of 1.10 when using Equation 6.3 for F_{ax}. The mean ratios confirm a perfect fit of the analytical model with the experimental values for the tests with five screws each (d = 5 mm and ℓ = 100 mm), even for significantly higher friction coefficients.

Table 6.2: Analytical load-carrying capacities $F_{V,exp}$ (in kN) and ratio $F_{V,test}$ / $F_{V,exp}$ of the analytical to the test values for Series 1 and 2.

Series	Surface	F_{ax} with Eq. 6.2	Ratio	F_{ax} with Eq. 6.3	Ratio
1	untreated DVW	33.0	1.23	34.1	1.18
	milled pyramids 1.0 mm	48.6	1.09	51.3	1.03
	milled pyramids 1.5 mm	52.8	1.01	55.4	0.96
	milled pyramids 2.0 mm	54.3	0.97	57.4	0.92
	milled circular	40.8	1.22	40.2	1.24
	sanded	42.5	1.18	43.9	1.15
	sandblasted	40.9	1.24	42.1	1.20
	coated with EpoxyTape (0.1 mm)	48.9	1.18	50.7	1.14
	coated with EpoxyTape (1.0 mm)	48.0	1.09	49.3	1.06
	coated with Griptape	34.6	1.13	35.4	1.10
2	milled pyramids 1.0 mm (offset)	50.0	0.97	52.1	0.94
	milled pyramids 1.5 mm (offset)	54.1	0.84	56.3	0.85

The smaller ratios determined for Series 2 are due to the smaller load-carrying capacities reached in the respective tests of Series 2. Due to a decreased inclination of the connector plates, less load was transferred through the direct contact and indention of the lower edge of the connector plate. Therefore, the test results were smaller (than for Series 1), while the analytical results predicted the same load-carrying capacities (as for Series 1).

Table 6.3 compares the tests of Series 3–5 and the tests with the different prototype connectors. The expected ultimate loads were now only calculated with the withdrawal capacity according to Equation 6.2 because the penetration length of the screws was greater than 140 mm. In addition, the tests of Series 3–5 were designed with the tensile capacity of the screws being decisive. For the tests of Series 3 and 5, the ratio of expected load to test load is almost equal to 1.0. For the tests of Series 4 with 15 screws and ℓ = 100 mm, the average ratio is significantly greater than 1.0, underestimating the load-carrying capacity by approximately 15%. The analytical model overestimates the load-carrying capacity by approximately 10% for the three connector prototypes, leading to average ratios smaller than 1.0. For the main beam to secondary beam tests with the connector prototype v3, the expected load is significantly higher than the test load and, therefore, the ratio is much smaller with only 0.79. That is because of different failure modes covered by the analytical model and that occurred during those tests. During the tests, tensile failure perpendicular to the grain occurred at the face grain end of the secondary beam. Consequently, the calculated results for this series are not considered in Figure 6.3 and for the calculation of the R^2-values.

Table 6.3: Calculated load capacities $F_{V,exp}$ (in kN) and comparison $F_{V,test}$ / $F_{V,exp}$ of the calculated values with the maximum loads from the tests of Series 3–5.

Series	Surface	Screw type/number		$F_{V,exp}$	$F_{V,test}$ / $F_{V,exp}$
3	milled circular	6x180	5	80.4	0.99
	milled pyramids 0.5 mm	6x180	5	82.2	1.03
4	untreated	6x100	15	87.8	1.36
	embossed	6x100	15	119	1.18
	milled pyramids 0.5 mm	6x100	15	132	1.16
5	embossed	6x200	15	191	0.97
6	milled pyramids 0.5 mm (v1)	6x200	10	164	0.91
	milled pyramids 1.0 mm (v2)	6x200	12	189	0.91
	beam-to-beam tests (v2)	6x200	12	176	0.98
7	milled pyramids 1.0 mm (v3)	8x300	20	518	0.94
	beam-to-beam tests (v3)	8x300	20	497	0.79

The evaluation of the test results showed that the displacements reached at ultimate loads were much higher than those reached at the maximal coefficient of friction. This difference leads to the assumption that the static friction coefficient may not be suited for the analytical model, leading to expected loads that are too high. Therefore, the input of the friction coefficient was changed to using the kinetic friction coefficient. This can also be explained mechanically: the load-carrying capacity determined in the tests was either the maximum load at failure of the connection (before reaching a displacement of 15 mm) or the maximum load at a displacement of 15 mm (when no failure occurred before). As the maximum load was evaluated at great displacements of 5–15 mm, depending on the connector, it also seems reasonably to use the kinetic coefficient of friction, also evaluated at great displacement of 10–15 mm. In Figure 6.3, the test loads are plotted against the expected load.

When using the static coefficient of friction for the analytical model, the lowest ratio value is 0.75, and the average ratio is 1.09. The R^2-value is quite high with 0.96. When using the kinetic coefficient of friction for the model, the lowest ratio value is 0.92, and the average ratio is 1.26, with an R^2-value of 0.92. Finally, when using the mean of static and kinetic coefficient of friction, the lowest ratio value is 0.85, and the average ratio is 1.17. R^2-value of 0.95. This shows that the static coefficient of friction can be used quite reliably to predict the load-carrying capacity for the herein presented connector configurations and failure modes.

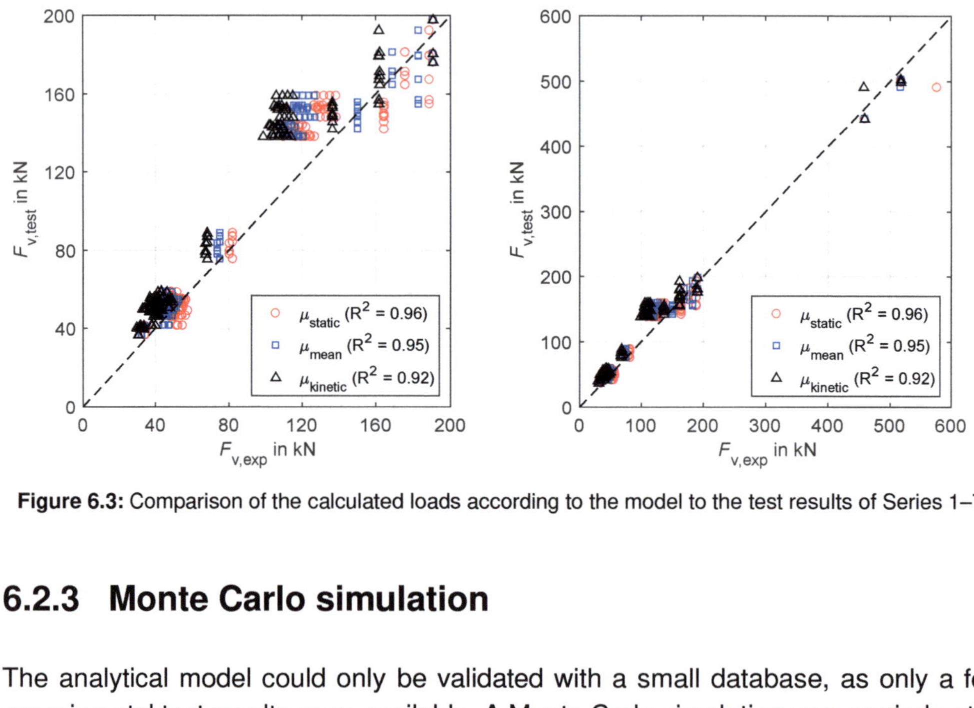

Figure 6.3: Comparison of the calculated loads according to the model to the test results of Series 1–7.

6.2.3 Monte Carlo simulation

The analytical model could only be validated with a small database, as only a few experimental test results were available. A Monte Carlo simulation was carried out to see how the predictive power of the model changes with a large number of results. Monte Carlo simulation is a powerful computational technique used in various scientific, engineering, and financial fields to model and analyse complex systems and processes. It relies on the generation of random samples to estimate mathematical results. These random input values follow specified probability distributions and can be used to evaluate the system's output. By incorporating randomness into the modelling process, valuable insights into the behaviour of the analytical model are given.

The most critical input value to simulate is the coefficient of friction. The other input parameters all correlate to some extent. For example, the density of the timber parts influences the withdrawal capacity of the screws, as well as the compressive strength perpendicular to the grain of the header. The friction coefficient, however, is uncorrelated and only influenced by the surface type and modification.

Only friction coefficients for the milled pyramid pattern were investigated for the Monte Carlo simulation. In Chapter 4, it was shown that the coefficients of friction follow a log-normal distribution. The mean values and standard deviations needed for the simulation were also taken from Chapter 4. Additionally, for the tests with prototype v3, the in-plane compressive strength of the DVW was simulated, using the data in [88] for the mean value and standard deviation. In total, 50 000 random values for the friction coefficient were simulated. Their logarithmic values are normally distributed, as shown in Figure 6.4. With the simulated friction coefficients and the simulated compressive strength of the DVW, load-carrying capacities according to Equation 6.6 were calculated.

The experimental test results for the tests with the connector prototype v1 were also tested for log-normal distribution using the Kolmogorov-Smirnov test. However, the sample size was relatively small, with $n = 9$. For the tests with prototypes v2 and v3, it was assumed that the experimental results were also log-normally distributed. Again, 50 000 values of test results for each connector prototype were simulated using the Monte Carlo method. The histogram in Figure 6.4 confirms the log-normal distribution of the simulated data.

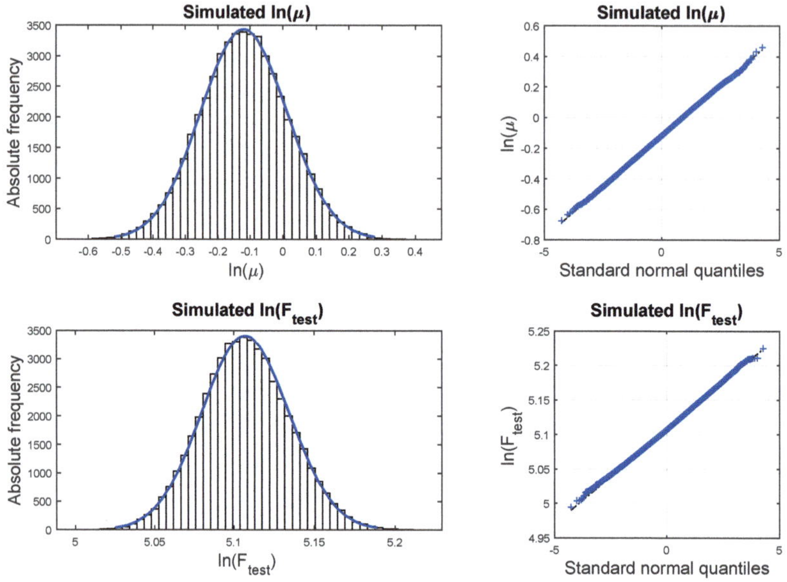

Figure 6.4: Log-normal distribution for simulated friction coefficients (top) and simulated test results (bottom).

Finally, the simulated expected values, $F_{V,exp,sim}$, were plotted against the simulated test results, $F_{V,test,sim}$. This was done for the static friction coefficient, the kinetic friction coefficient, and the mean friction coefficient. The results are shown in Figure 6.5. The influence of the coefficient of friction on the load-carrying capacity can be seen in the different diagrams. The results shift from the right to the left with decreasing friction coefficients. The dashed line stands for a ratio of F_{test}/F_{exp} of 1.0. If the values are above the dashed line, the experimental test results are higher than the analytically expected results, and the model is on the safe side. Hence, if the values are below the dashed line, the expected results are higher than the test results, and the model overestimates the load-carrying capacity. The lowest ratio for the simulation with μ_{static} is approximately 0.6, and the mean ratio is 0.9. For the simulation with μ_{mean}, the lowest ratio is approximately 0.7 and the mean ratio is 1.0. For the simulation with $\mu_{kinetic}$, the lowest ratio is approximately 0.8, and the mean ratio is 1.1. The R^2-value for each of the different friction coefficients is also given in the diagrams. Each R^2-value was calculated concerning the dashed line. All three R^2-values are very close, with 0.93 for the analytical model using the static coefficient of friction and 0.94 for the model using either the kinetic or the mean coefficient of friction.

The shape of the point cloud of the data of prototype v3, and especially the extensive range of the expected data on the x-axis, confirms that the static coefficient of friction is, in most cases, too high to use reliably for the analytical calculation of the load-carrying capacity. Combined with a high coefficient of variance (COV) for the static friction coefficient of 13%, the expected data ranges from about 300 to 700 kN. The impact of a smaller kinetic friction coefficient can be seen in the corresponding point cloud. Although the COV-value increases to 16%, the expected data range decreases to about 300 to 600 kN. The extremely small range of the simulated test data on the y-axis is mainly due to the extremely small COV values for the experimental test data of prototype v3, i.e. 1% with $n = 4$.

In conclusion, the Monte Carlo simulation confirmed the assumptions made for the analytical model by simulating 50 000 values. Additionally, it was confirmed that the static coefficient of friction is, in most cases, too high to use reliably for the analytical calculation of the load-carrying capacity.

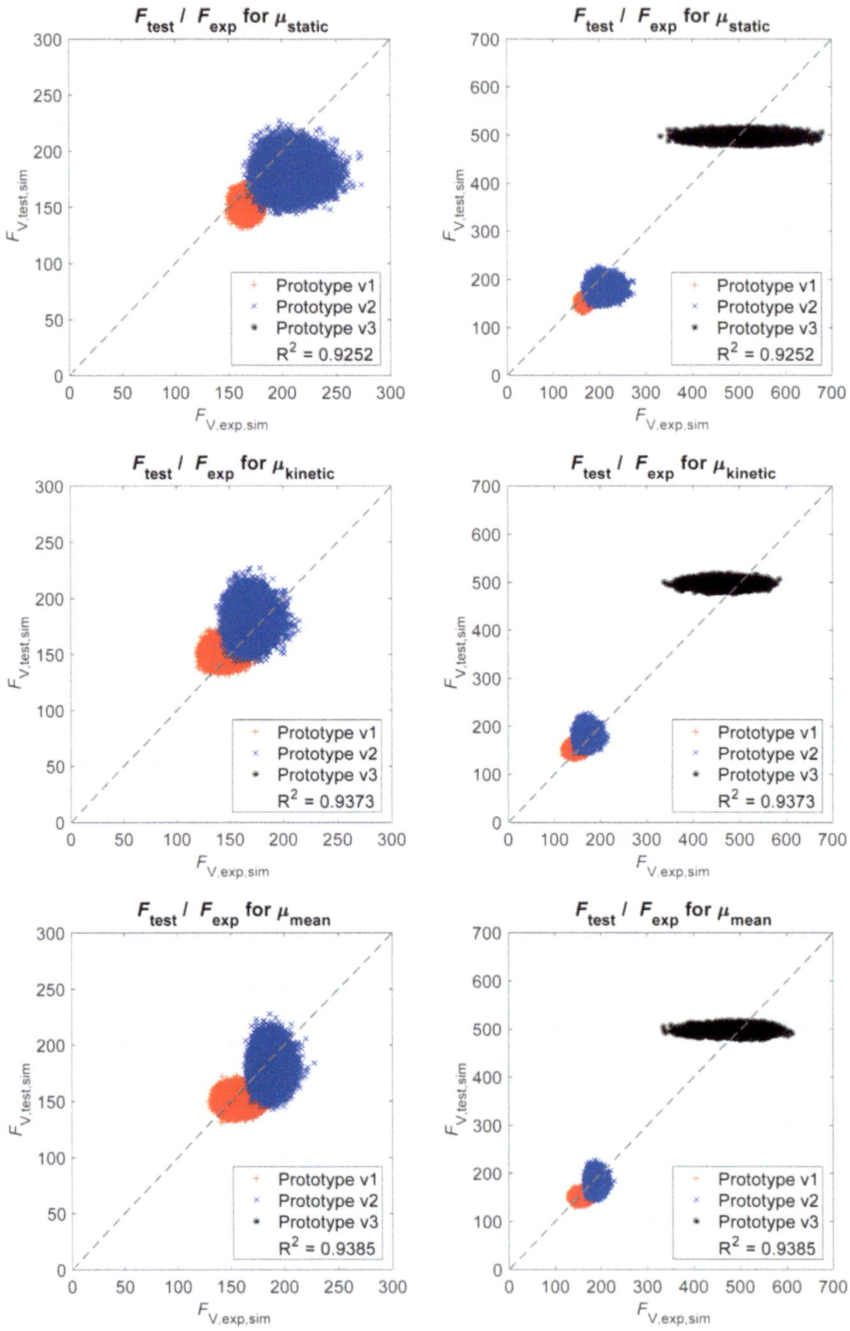

Figure 6.5: Ratios of F_{test} to F_{exp} for Monte Carlo simulation and different friction coefficients.

6.2.4 Comparison with characteristic values

With the Monte Carlo simulation, 50 000 values of test results for each connector prototype were simulated, i.e. their load-carrying capacity. Characteristic values (5% quantiles) of the respective load-carrying capacities were determined according to EN 14358. The statistical value $k_s(n)$ was 1.76. The expected load-carrying capacity was calculated with Equation 6.6. For the values of F_{ax} and F_{tens} the characteristic values according to the screw's ETA were used. For the compressive strength of the DVW connectors and the timber, also characteristic values were used. For the coefficient of friction, different values were investigated and compared to each other. On the one hand it was differentiated between the static and the kinetic friction coefficient, on the other it was now also differentiated between the respective mean value and characteristic value.

The ratios of $F_{test,k}$ to $F_{exp,k}$ are plotted in Figure 6.6. In red, the results for the expected load calculated with static friction coefficients, and in grey, the results for the expected load calculated with kinetic friction coefficients. Figure 6.6a shows that for the small connectors all test results are higher than the expected load. Figure 6.6b shows that only the high results for the connector prototype v3 are overestimated. This leads to the conclusion that the characteristic load-carrying capacity of a connection with inclined screws and increased friction in the shear plane can be predicted reliably when using the mean value of the static coefficient of friction. When using the mean value of the kinetic coefficient of friction, the results are more conservative and on the safe side.

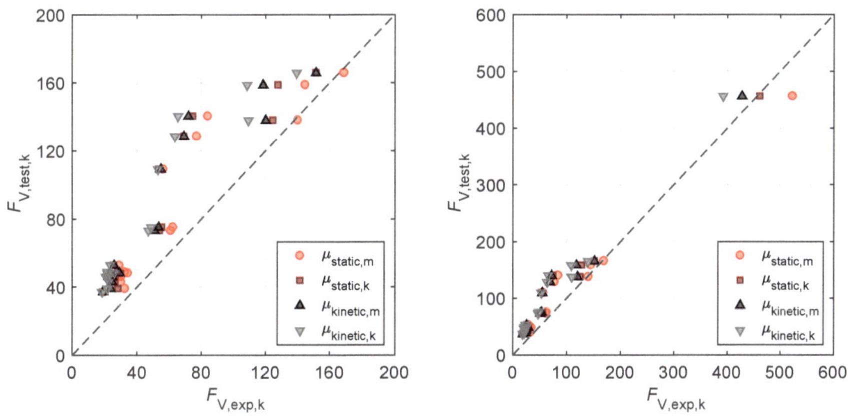

Figure 6.6: Ratio of characteristic load $F_{test,k}$ to $F_{exp,k}$ for different friction coefficients (mean values vs. characteristic values).

6.2.5 Influence of the insertion angle

With the proposed model, it is possible to investigate the influence of the insertion angle ε on the load-carrying capacity and highlight the additional part due to friction. The black solid line in Figure 6.7a shows the part parallel to the shear plane of the axial load-carrying capacity due to the inclination of the screws (i.e. $F_{ax} \cdot \cos\varepsilon$). For an insertion angle of 0°, the theoretical total load-carrying capacity is 100% of the axial load-carrying capacity of the screw (as all the load parallel to the shear plane would be transferred by the axial capacity of the fastener). For an insertion angle of 90°, the load-carrying capacity is 0% of the axial capacity of the screw (as all the load parallel to the shear plane would be transferred by the embedding capacity of the fastener). The coloured dashed lines show the part parallel to the shear plane of the load-carrying capacity due to friction (i.e. $F_{ax} \cdot \mu \cdot \sin\varepsilon$). The curves show an increase in the load-carrying capacity with an increasing insertion angle. The different colours indicate different friction coefficients. It shows that the portion due to friction increases with increasing insertion angle. The coloured solid lines represent the total load-carrying capacity for the different friction coefficients, i.e. the sum of the parallel part due to the inclination and the additional parallel part due to friction. The evaluation shows that for smaller friction coefficients, smaller insertion angles are favourable, whereas for greater friction coefficients an optimum is reached at an insertion angle of 45°.

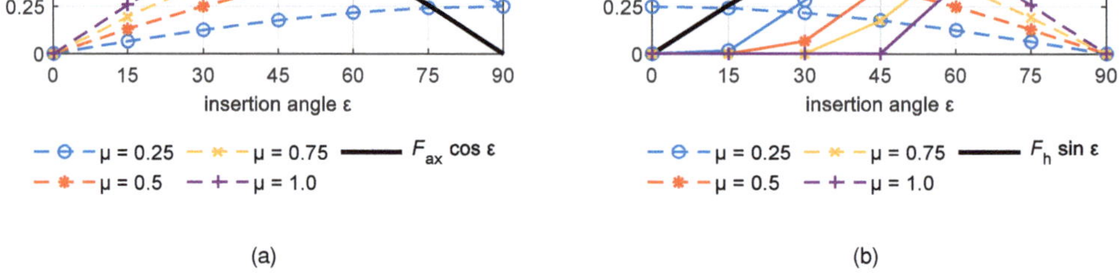

Figure 6.7: Influence of insertion angle and friction coefficient on axial load-carrying capacity (a) and lateral load-carrying capacity (b).

If the embedding strength of the fastener is considered, the load-carrying capacity is calculated according to Equation 6.7. The derivation of the equation is presented in detail in [7].

$$F_{\mathrm{V}} = F_{\mathrm{ax}} \cdot (\cos\varepsilon + \mu \cdot \sin\varepsilon) + F_{\mathrm{h}} \cdot (\sin\varepsilon - \mu \cdot \cos\varepsilon) \tag{6.7}$$

with F_V load-carrying capacity of the connection parallel to the shear plane

F_{ax} axial capacity of the fastener

F_h embedment strength of the fastener

ε insertion angle

μ friction coefficient

Figure 6.7b now shows the influence of the insertion angle and the friction coefficient on the lateral load-carrying capacity. The black solid line shows the lateral part of the load-carrying capacity due to the insertion angle (i.e. $F_h{\cdot}\sin\varepsilon$). At an insertion angle of 90°, the load-carrying capacity is 100% of the lateral load-carrying capacity of the fastener (as all the load parallel to the shear plane would be transferred by the embedding capacity of the fastener). At an insertion angle of 0°, the load-carrying capacity is 0% of the lateral load-carrying capacity of the fastener (as all the load parallel to the shear plane would be transferred by the axial capacity of the fastener). Again, the dashed coloured lines show the additional part of the load-carrying capacity due to friction (i.e. $F_h{\cdot}\mu{\cdot}\cos\varepsilon$) and the solid coloured lines show the total lateral load-carrying capacity. Due to equilibrium reasons, the portion of the load due to friction is subtracted from the total lateral capacity.

According to Equation 6.7, the total load-carrying capacity for timber-to-timber connections is the sum of the results of the two figures (solid coloured lines). However, the values evaluated in the two figures cannot simply be summed up, as the axial and lateral capacities of dowel-type fasteners are not in the same order of magnitude. Therefore, in the design of connections with inclined screws, the embedding strength of the screws is disregarded. The evaluation of the influence of the insertion angle shows that depending on the friction coefficient, insertion angles of 15–45° are favourable. An additional portion due to embedment is only visible for insertion angles greater 30–45°. Overall, it shows that an insertion angle of 45° is a good choice for connections with inclined screws and increased friction in the shear plane.

Figure 6.8 shows the evaluation of Equation 6.7 with the actual test results of the connection tests. The part due to embedment was neglected. The normal force occurring in the screw was analysed by solving Equation 6.7 according to F_{ax}. The values are plotted against the insertion angle ε. The tensile capacity of the screws used was taken as the normal force at an angle of 0°. This also results in the subdivision of the data according to the used diameters. For an angle of 45°, Equation 6.7 was converted to the normal force F_{ax} and the ultimate load from the tests was used for F_V. For an angle of 90°, a normal force of 0 kN is assumed in the screws. The data evaluation shows a trend similar to that seen in Figure 6.7a.

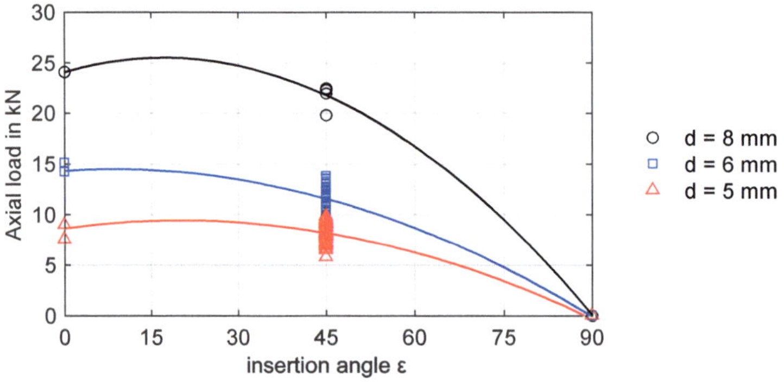

Figure 6.8: Axial load per screw for different insertion angles.

It can be seen that the load-carrying capacity of a connection can be increased in different ways. On the one hand, the coefficient of friction and, thus, the additional part of the overall load-carrying capacity due to friction can be increased. In order to activate friction in the shear plane, a contact pressure perpendicular to the shear plane is required. This contact pressure can also be generated in various ways. Either by inclining the fasteners and thus due to equilibrium reasons. Alternatively, by increasing the normal force in the fastener. In the case of inclined fasteners, the normal forces increase automatically due to the insertion angle. In the case of fasteners inserted perpendicular to the shear plane, large displacements are necessary to increase the normal force in the fastener. Therefore, for fasteners inserted perpendicular to the shear plane, pre-tensioning is a feasible method. This was demonstrated in the tests with the single-shear steel-to-timber connections with bolts. Here, however, also the stiffness of the used materials is of great importance.

6.3 2D FE-Model

6.3.1 Structure of model

In order to be able to make statements about the deformation of the connection, the analytical model was extended to a numerical two-dimensional model. The basis for the two-dimensional framework model follows the approach in [89]. The connection is modelled as a combination of *beam elements* and *spring elements*. Three types of springs are used and modelled with Abaqus® [121]:

- *Tensile springs* to model the tensile behaviour of axially loaded screws
- *Embedment springs* to model the embedment behaviour of laterally loaded screws
- *Contact springs* to model the interaction between the connector and timber beam

The springs were modelled as two-dimensional *connector elements* CONN2D2. The springs representing the screws had the properties of the tensile springs (Section 6.3.2.1) in their longitudinal direction and the properties of the embedment springs (Section 6.3.2.2) perpendicular to their axis. The contact between the timber and connector plate was modelled with the contact springs. The properties in their longitudinal direction matched the compressive properties (Section 6.3.2.3), depending on the grain direction of the timber. The properties perpendicular to their axis were defined by the coefficient of friction determined in Chapter 4. The densified veneer wood connector plates were modelled as *beam elements* (B21 – 2-node linear beam). The material properties were taken from test results determined in [88]. The timber parts were modelled as *rigid elements* (R2D2 – 2-node linear rigid link) without further material properties. The exact input parameters are in Appendix A.4, Table A.10–A.14. Figure 6.9 shows the model with its boundary conditions and the placement of the different springs.

The load is applied by displacing the respective boundary condition by 15 mm in the y-direction. The reaction force required for this displacement is evaluated as the total load-carrying capacity of the connector. The corresponding displacement is calculated as the relative displacement between the main and secondary beam, each at half of the connector's height, highlighted in Figure 6.9 with red dots. Also highlighted with red dots are the locations where the displacements in the x-direction of the secondary beam were measured. With this horizontal displacement, the rotation of the secondary beam was determined.

The main beam is constrained in x- and y-direction. Deformations due to the material's properties are depicted by the springs (i.e. compression perpendicular to the grain) and are therefore not constrained by the boundary conditions. The secondary beam is not constrained with boundary conditions to allow for vertical and horizontal displacement. In the first step, only a vertical load is applied at the top of the secondary beam, and no bending moment.

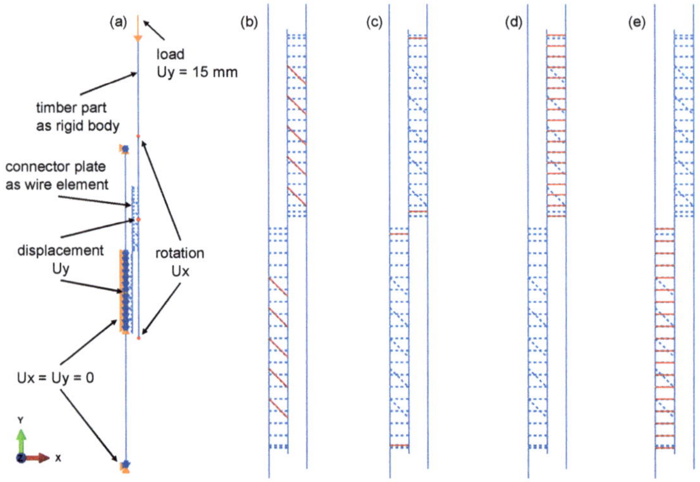

Figure 6.9: Main beam to secondary beam model with tensile and embedment springs (credit to [90]): (a) boundary conditions; (b) inclined screws; (c) perpendicular screws; (d) contact springs for compression parallel to grain and (e) perpendicular to grain.

6.3.2 Characteristics of springs

6.3.2.1 Tensile springs

Axial stiffness

Tests to determine the axial stiffness of fully threaded screws were performed in [57]. Blaß and Steige compared three different testing methods (shear tests with inclined screws, shear tests with cross-wise arranged screws and withdrawal tests) to determine the stiffness values. With the results, an equation to calculate the axial stiffness was derived. In shear tests with two timber parts connected by inclined screws, the axial stiffness is measured, considering friction between the two timber parts. In shear tests with cross-wise arranged screws, the influence of friction is excluded. Also, withdrawal tests allow for a friction-free determination of the axial stiffness. A parameter analysis was carried out to investigate the effect of various influencing parameters on the stiffness values. The penetration depth ℓ_{ef}, the outer thread diameter d and the angle α between the grain direction and screw axis were examined. The results showed that shear tests with cross-wise arranged screws were a suitable test method, and the determined axial stiffness values could also be used for connections with inclined screws (which were not cross-wise arranged). To calculate K_{ax}, Equation 6.8 was determined:

$$K_{\mathrm{ax}} = 0.48 \cdot d^{0.4} \cdot l_{\mathrm{ef}}^{0.4} \cdot \rho_{\mathrm{m}}^{0.3} \cdot \alpha^{0} \tag{6.8}$$

with	d	diameter of the screw [mm]
	ℓ_{ef}	effective penetration length of the screw [mm]
	ρ_m	mean density [kg/m^3]
	α	insertion angle [°]

From the withdrawal tests, it is evident that the screws show plastic deformation. However, information on determining the elongation at the ultimate load is missing in [57]. Therefore, it is assumed in a highly simplified manner that a minimum elongation of 1.0 mm is reached at the ultimate load. The displacement behaviour is described with three linear sections: For the first linear elastic section, the axial stiffness, according to Equation 6.8, is applied up to 0.8 times the withdrawal capacity. The remaining curve part is divided halfway to the ultimate displacement at 0.95 times the withdrawal capacity. The overall behaviour of the spring is described as nonlinear elastic.

Axial load-carrying capacity

The same equations as for the analytical model apply to the withdrawal capacity. However, to make optimum use of the connection, the penetration depth of the screws should be chosen to such an extent that their tensile load-carrying capacity becomes decisive. The data from the respective ETA (characteristic values only) can be used for the tensile strength, or tests according to EN 14592 [113] can be performed.

Figure 6.10 shows examples of load-displacement behaviours for the described tensile spring, depending on the screws' penetration depths. The spring fails when reaching the ultimate load at a displacement of 1 mm. As per definition, the spring can only be subjected to tensile forces.

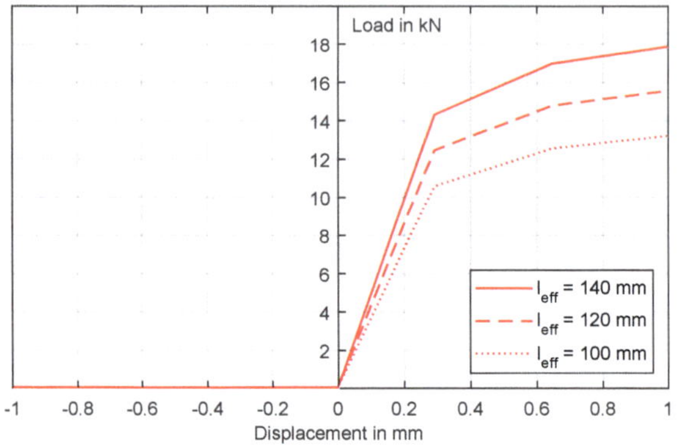

Figure 6.10: Axial load-displacement curves for tensile springs with varying penetration depths ℓ_{eff}.

6.3.2.2 Embedment springs

Lateral stiffness

In general, the load-carrying capacity and stiffness values of screws only consider either load parallel or perpendicular to the grain. Jockwer et al. [91] investigated the behaviour of fully threaded screws under a combined load. This load combination occurs, for example, in system connectors with inclined screws when the connector is submitted to both shear forces and a bending moment. The load-carrying capacity and stiffness of inclined screws, which are loaded perpendicular to the grain direction, are lower than compared to vertically inserted screws. The load causes large deformations of the screws and failure perpendicular to the grain of the timber. The load-carrying behaviour of such a connection can be divided into one component acting in the longitudinal direction of the screw and one component acting lateral to the screw's axis. The axial component F_{ax} is transmitted by the bond between the thread and the timber. The lateral component F_{v} must be absorbed by the embedment capacity of the timber. At the surface of the timber, the embedment stresses are close to zero and lead to the splitting of the timber. Starting at a depth of x_1, the timber's embedment strength takes up the load. The depth x_1 can be determined by a force equilibrium, considering the embedment strength f_{h}, the effective diameter d_{ef}, the rolling shear strength $f_{\mathrm{v,roll}}$ and the screw's inclination angle γ (see Figure 6.11a). For simplification, it is assumed that the embedment strength is zero up to the depth x_1 and that the full embedment strength is activated afterwards. Figure 6.11b shows the influence of the angle between the screw axis and grain direction and the relationship between the axial stiffness and the lateral stiffness of the screw. At an angle of 90°, the total axial stiffness is available while only 10% of the lateral stiffness. Therefore, a stiffness of 10% of K_{ax} is assumed for the embedment springs.

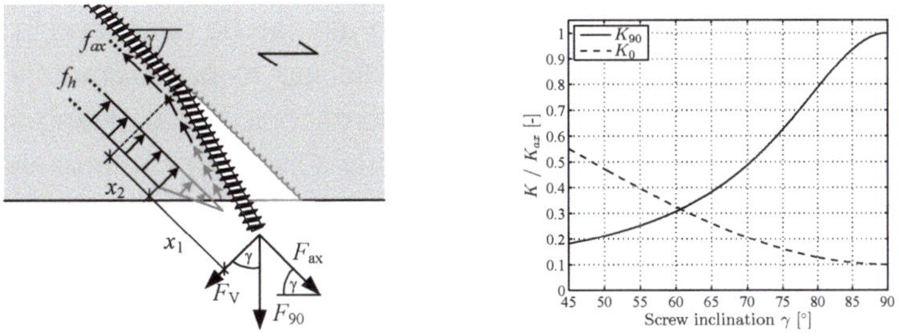

(a) (b)

Figure 6.11: Stresses acting in a screwed connection when loaded perpendicular to the grain (a) and impact of screw axis to grain angle γ on the relative stiffness K/K_{ax} (b) [91].

Lateral load-carrying capacity

Analogous to the lateral stiffness, the lateral load-carrying capacity is assumed to be 10% of the axial load-carrying capacity. Figure 6.12 shows load-displacement behaviours for the described embedment spring, with the same penetration depths as before. Unlike the tensile spring, the embedment spring can be loaded in both directions. In addition, at a displacement of 1 mm, the ultimate load of the spring is reached. However, the spring can further increase the displacement without increasing the load because of the ductile behaviour of the embedment strength.

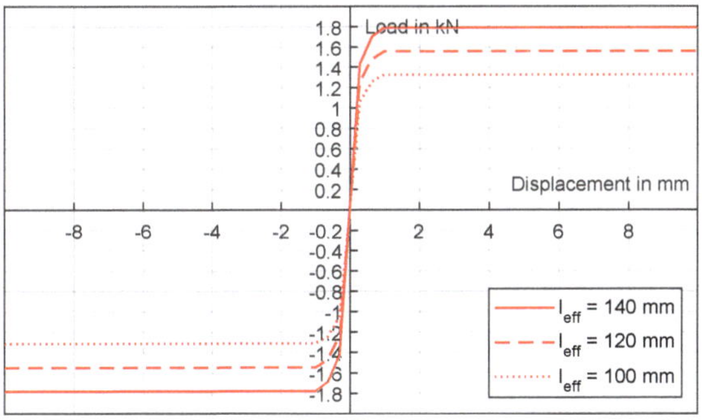

Figure 6.12: Lateral load-displacement curves for tensile springs with varying penetration depths ℓ_{eff}.

6.3.2.3 Contact springs

Load-carrying capacity parallel to the grain

The loading behaviour of timber parallel to the grain can be described in three stages. In the first stage, the behaviour is linear-elastic, up to approximately 70% of the compressive strength. In the second stage, microscopic shear/sliding surfaces form in the fibres, and the timber behaves non-linearly. Once the compressive strength is reached, kink bands form, and the compressive strength drops to 85%. Further loading leads to increasing deformations at a relatively constant residual load-carrying capacity.

Only the first stage, with its linear-elastic behaviour, is modelled for the spring's properties, and the results are checked to be less than 70% of the compressive strength. Equation 6.9 is used to determine the spring's properties. Assuming a mean compressive strength parallel to the grain of 36 N/mm² and a modulus of elasticity of 11 500 N/mm² for glulam GL 24h, this results in a limiting strain of ε = 0.003.

$$\sigma_{\mathrm{c},0} = E \cdot \varepsilon = E \cdot \frac{\delta l}{l_0} = \frac{F_{\mathrm{c},0}}{A} \tag{6.9}$$

By dividing the loaded area (i.e. the area beneath the connector plate) into smaller parts, the maximum force per spring can be calculated with Equation 6.9. To avoid tensile forces in the compression spring, the displacement at 0.1 N tensile force is set to 100 mm, practically leading the spring to fail.

Load-carrying capacity perpendicular to the grain

In general, the loading behaviour of timber perpendicular to the grain is elastic-plastic. A compressive load perpendicular to the grain leads to plastic deformations without a load drop after exceeding an elastic limit. In [92] a model is presented to calculate the load-carrying capacity perpendicular to the grain as a function of the material, the possible load propagation (one-sided or two-sided) and permitted indentation. The parameter $k_{c,90}$ is firstly defined as a function of the indentation u in the timber, see Equation 6.10:

$$k_{\mathrm{c},90}(u) = k_{\mathrm{a}} \cdot (1 - e^{-k_{\mathrm{b}} \cdot u}) \tag{6.10}$$

The values for k_a and k_b depend on the timber material used, the loading direction, and the possible load propagation. The transmittable force $F_{c,90,R}$ is calculated with Equation 6.11 by considering the width b of the loaded surface perpendicular to the grain, the length ℓ of the loaded surface, a length ℓ_{dis}, and the compressive strength perpendicular to the grain $f_{c,90}$ of the material. The effective length increases linearly until it reaches ℓ_{dis} at an indentation of u = 15 mm.

$$F_{\mathrm{c,90,R}}(u) = b \cdot (l \cdot k_{\mathrm{c,90}}(u) + l_{\mathrm{dis}}) \cdot f_{\mathrm{c,90}} \tag{6.11}$$

By substituting Equation 6.10 into Equation 6.11, a direct correlation is obtained between indentation in the wood and the resulting reaction force. Using this approach, the load-carrying capacity of timber can be modelled as a non-linear spring. Values for k_a, k_b and ℓ_{dis} are given in [92]. For the compressive strength perpendicular to the grain $f_{c,90}$ = 4.3 N/mm^2 was assumed.

$$F_{\mathrm{c,90,R}}(u) = b \cdot (l \cdot 1.5 \cdot (1 - e^{-0.4 \cdot u}) + l_{\mathrm{dis}}) \cdot f_{\mathrm{c,90}} \tag{6.12}$$

Frictional forces

As described in Chapter 2, the part of the load transmitted by friction corresponds to the frictional force, which is the product of the force perpendicular to the friction surface and the coefficient of friction μ (see Equation 6.13). The force perpendicular to the friction surface or the shear plane is defined by the axial spring force of the contact spring. Multiplied by the coefficient of friction, the result is the frictional force. For this, the contact springs are assigned a coefficient of friction, which accordingly limits the load transfer perpendicular to the spring axis to μ times the axial load of the spring.

$$F_{\mathrm{f}} = \mu \cdot F_{\mathrm{n}} \tag{6.13}$$

Figure 6.13 shows the properties of the contact spring for loads perpendicular to the grain. When reaching the ultimate compressive load, the displacement will continue to increase. The dashed line shows the load-carrying capacity perpendicular to the longitudinal axis, reduced by the friction coefficient μ. The algebraic signs are adjusted accordingly in the model, with compressive forces being negative and tensile forces being positive.

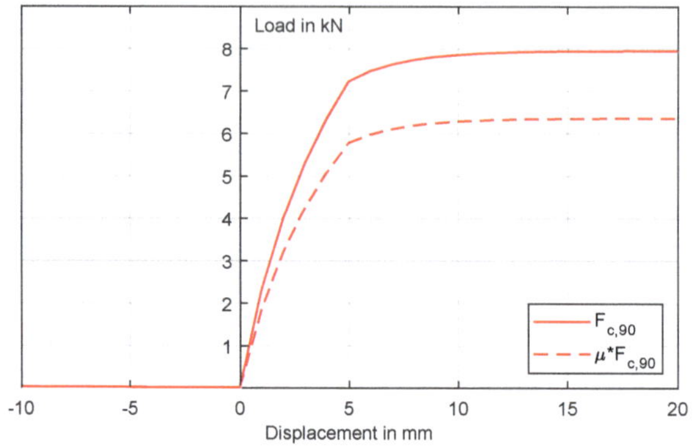

Figure 6.13: Load-displacement curves for contact springs in longitudinal direction (solid line) and perpendicular to the axis (dashed line).

Rotational capacity of connector

The rotational capacity of the connector is modelled with a pair of horizontal springs. The two springs have a leverage arm that is half the insertion nozzle of the connector plate. The rotational moment is divided into a pair of forces and assigned to two horizontal springs. The rotational capacity of both connector prototypes was determined in [3].

6.3.3 Validation of springs

In order to check whether the given formulas for the calculation of the properties of the springs are applicable and whether the 2D model with the three different types of springs works, shear tests with inclined screws performed by Blaß & Steige [57] were simulated. The tested experimental setup is shown in Figure 6.14a and the corresponding spring model in Figure 6.14b. The two screws were inclined by 45° and are modelled as springs with a maximum load-carrying capacity in their axial direction according to Equation 6.2 and with an axial stiffness according to Equation 6.8. Reaching the ultimate load-carrying capacity leads to the failure of the springs. Likewise, no compressive forces can be absorbed. The load-carrying capacity and stiffness under load perpendicular to the axis are 10% of the corresponding values in their axial direction, according to [91]. The contact between the two timber parts is modelled with springs with the compressive properties of wood. In the longitudinal direction, the load-carrying capacity is perpendicular to the grain according to Equation 6.12. Perpendicular to the spring's axis, the load-carrying capacity is the frictional force in the shear plane, i.e. μ times the force in the axial direction of the spring (following the laws of friction and Equation 6.13). The load-displacement curves from the experimental tests and the model are shown in Figure 6.15. It shows that the assumptions and simplifications reflect the actual load-displacement behaviour well. The average load reached in the numerical calculation is 36.5 kN and only 2% above the experimental results. The deviation for the stiffness values K_i and K_s is 13% and 2%. Minor variations in the results can be made by either changing the coefficient of friction (in [57], friction coefficients between 0.4–0.5 were recorded for these tests) or changing the withdrawal capacity of the screws by using Equation 6.3 instead of Equation 6.2.

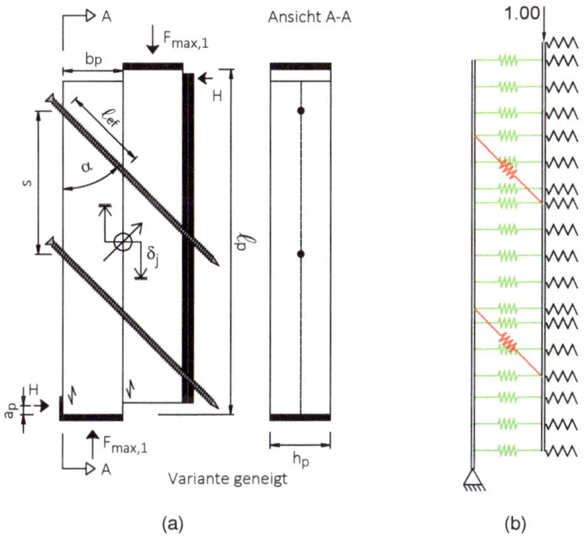

Figure 6.14: (a) Test setup for push-out test with inclined screws [57] and (b) 2D model to validate the spring's settings (according to [93]).

Thus, it could be shown that with the spring model and the three types of springs, connections with inclined screws, which are axially loaded, can be simulated. Both the load-carrying capacity and the stiffness, and thus the deformation of the connection, can be calculated without significant deviations.

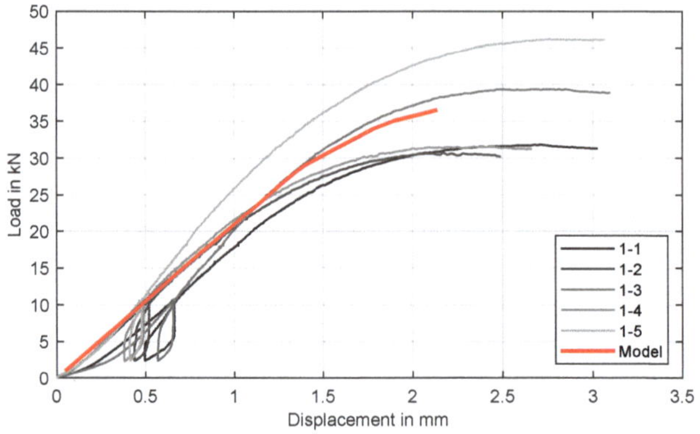

Figure 6.15: Comparison of 2D model with springs and push-out tests with inclined screws.

6.3.4 Comparison with experimental results

The results from the model are compared to main beam to secondary beam tests carried out in Chapter 5, Section 5.3.4. For the connector prototype v2, five tests were carried out in total. Figure 6.16 shows the corresponding load-displacement curves. The prediction from the model fits very well with the experimental test results. An ultimate load of 158 kN is reached before the FE model stops due to convergence problems. For the tests, the mean load was 172 kN, resulting in a deviation of 8%. The deviation of the stiffness values is between 14–22% for K_i and K_s.

In Figure 6.18 and Figure 6.19, the loads in the springs are given. The respective maximum spring force is shown in dashed lines, i.e. 14.1 kN for the tensile load-carrying capacity of the screws and 10% of this, i.e. 1.41 kN for the embedment capacity. The results for one screw per row are shown. The numbering of the rows is from the top edge to the bottom edge of each connector. The inclined screws in the secondary beam (the side where the load is applied) are all loaded in tension, but the tensile load-carrying capacity is not reached for any screw. The two upper rows of screws are loaded more than the lowest row of screws.
On the main beam side, there are also different loads depending on the position of the screws. The topmost screws are also loaded the most. The tensile load-carrying capacity is reached, but this does not lead to failure of the screws. The force in the spring does not increase with increasing deformation, and the failure criterion is limited to the forces and not the spring's deformation. As can be seen in Figure 6.17a, this also occurred during some of the tests, where screws were withdrawn, although analytically, the tensile capacity was decisive. The lowest row of screws on the main beam is already

loaded in compression after a short time due to the twisting of the connector plates and consequently does not transmit any further forces. This failure mode was also observed during the tests, as shown in Figure 6.17b. The head of the perpendicular screws sticks out; with the inclined screws, it can only be guessed that they are not in direct contact with the connector. Perpendicular to the spring axis, all screws reach the embedment strength.

The evaluation of the contact springs shows large compression forces perpendicular to the grain in the main beam underneath the connector plate. The load almost reaches the compressive capacity, especially at the bottom end of the connector plate, whereas the springs at the top are only loaded to half. This twisting behaviour was also observed during the tests, as shown in Figure 6.17c. At the end grain end of the secondary beam, compressive forces occur parallel to the grain direction but are nowhere close to the compressive capacity. In addition, the compressive load is below 70% of the load-carrying capacity, and thus, the assumption of linear-elastic compressive behaviour parallel to the grain can be confirmed.

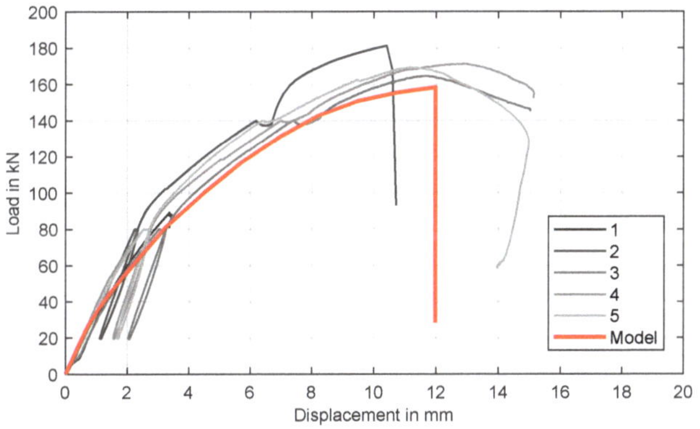

Figure 6.16: Comparison of 2D model with springs and full-scale tests of connector prototype v2 with increased friction.

The evaluation of the rotation of the connector, however, shows a significant discrepancy. Figure 6.20 shows the load-rotation curves for the experimental tests (grey lines). The rotation does not exceed 0.4°. The red line shows the load-rotation curve for the 2D model. The secondary beam rotates by 16° before reaching the ultimate load. First, it was believed to be due to missing contact springs in the model. No contact springs are defined between the lower connector part and the secondary beam, and no springs are defined between the secondary and the main beam. However, as the experimental

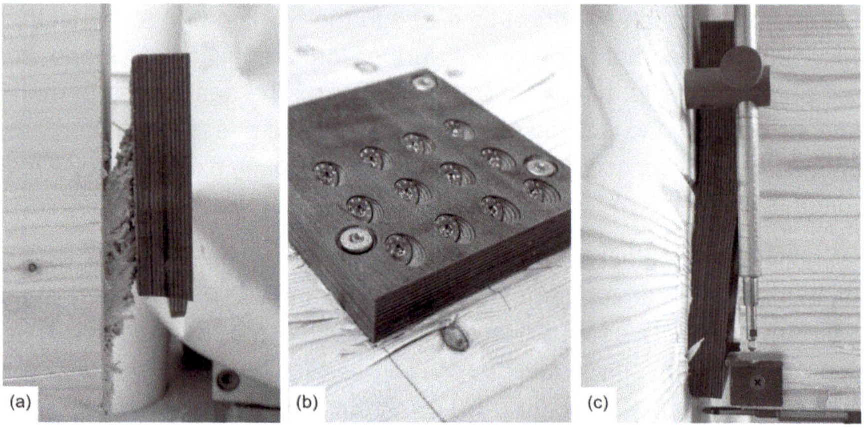

Figure 6.17: Connections after tests: (a) reaching of withdrawal capacity; (b) push-out of screws; (c) failure perpendicular to the grain.

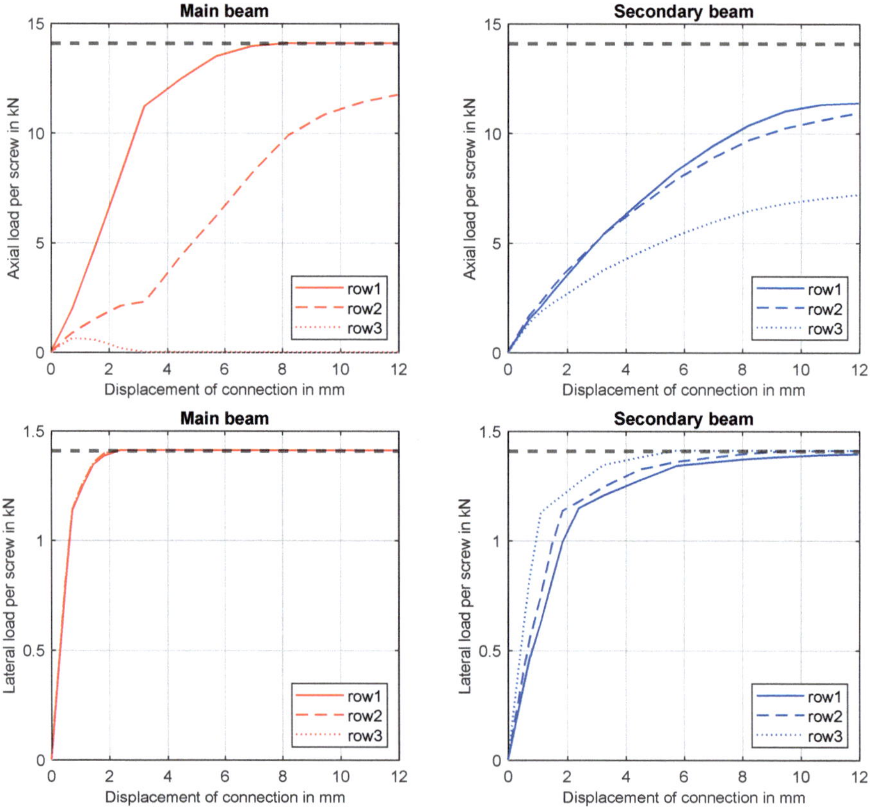

Figure 6.18: Spring loads (screw properties) of model with connector prototype v2.

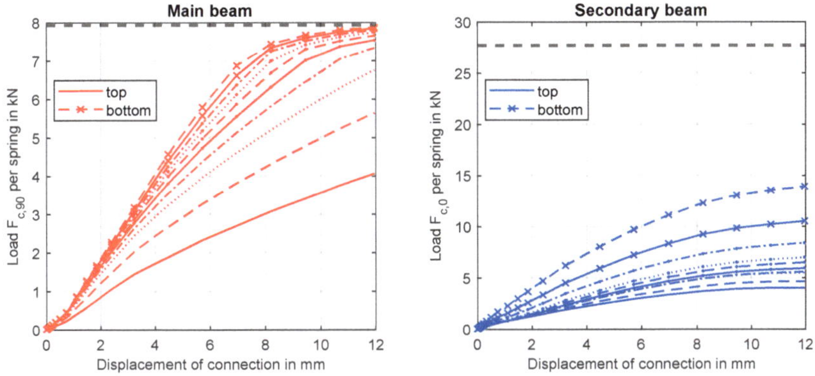

Figure 6.19: Spring loads (timber properties) of model with connector prototype v2.

tests showed, there was almost no contact between the bottom connector plate and the end grain of the main beam. Additionally, there was no direct contact between the main and the secondary beam. So, missing contact definitions cannot be the explanation.

Another possible explanation could be the properties of the compression springs perpendicular to the grain. Their load-displacement curve allows for large displacements of up to 50 mm in their longitudinal direction. However, the evaluation of the deformation of the compression springs perpendicular to the grain showed maximum displacements of only 15 mm. That is still quite large and could not be confirmed to that extent with the experimental tests, which showed a significant impression of the bottom connector.

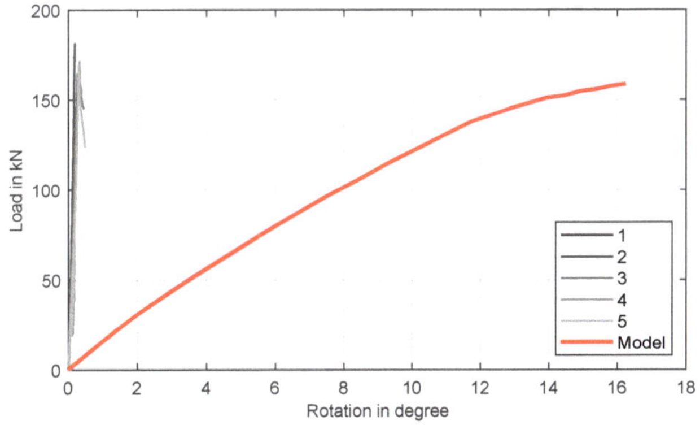

Figure 6.20: Rotation of 2D model with springs (red line) compared to full-scale tests of connector prototype v2 with increased friction (grey lines).

In the final step, the model was extended to the actual geometry as experimentally tested. The model is shown in Figure 6.21a. The load is applied at the top of the secondary beam at a distance from the shear plane, allowing for an additional bending moment due to the eccentricity of the shear plane. The secondary beam is constrained in the middle at two points in the x-direction. The load-displacement curve in Figure 6.21b shows the result from the model compared to the tests. The model reached an ultimate load of 208 kN, overestimating the experiments by 21%. The stiffness also increased. Here, the deviation was 27–46% for K_s and K_i respectively.

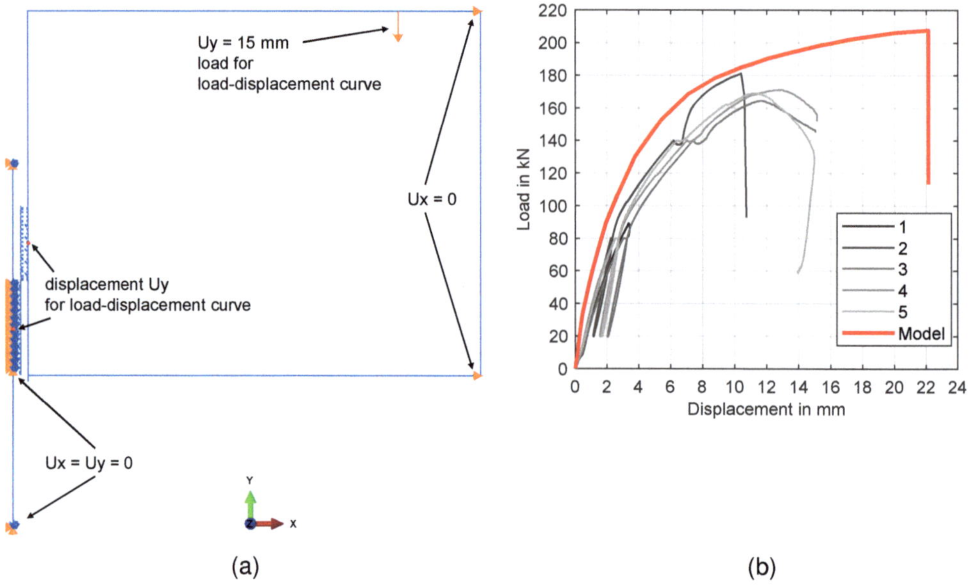

Figure 6.21: 2D model with springs of beam-to-beam tests (a) and comparison with results of tests with connector prototype v2 (b).

For the connector prototype v3, three tests were carried out in total. The corresponding load-displacement curves are shown in Figure 6.22 (grey lines). The first model (Model-1) was analogous to the model shown in Figure 6.9 with the load applied on top of the secondary beam. Again, no boundary conditions were set for the secondary beam. The prediction from the model fits very well with the experimental test results (red line). An ultimate load of 385 kN is reached before the FE model stops due to convergence problems. For the experimental tests, the mean load was 394 kN, resulting in a deviation of merely 2%. The stiffness of the model is lower than in the experiments. The deviation is between 26–28% for K_s and K_i. However, the rotation of the connector was overestimated to a great extent; see Figure 6.23 (red line). At the ultimate load, the rotation is about 10°, a multiple of the actual rotation recorded in the experiments. The same thought process applies, and there is no real explanation for the extremely high rotation.

Again, the model was extended in a second step to allow an additional bending moment (see Figure 6.21a). The secondary beam was constrained in the x-direction at two corners. The resulting load-displacement curve is given in blue (Model-2) in Figure 6.22. As before, the load can be further increased, and the experimental results are overestimated. The ultimate load was 456 kN, 16% higher than the tests. Additionally, the stiffness values increased and overestimated the experiments by 40% for both K_i and K_s.

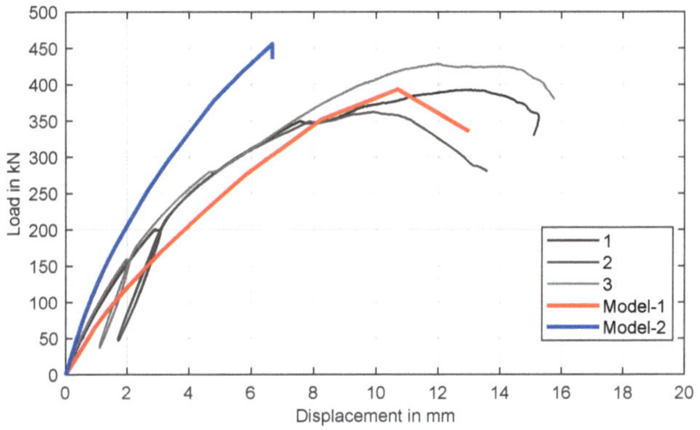

Figure 6.22: Comparison of 2D model with springs and full-scale tests of connector prototype v3.

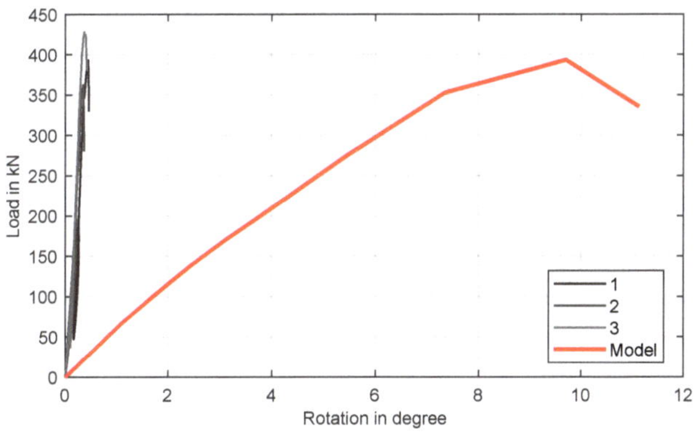

Figure 6.23: Rotation of 2D model with springs compared to full-scale tests of connector prototype v3.

The spring loads are given in Figure 6.25. The respective maximum spring force is shown in dashed lines, i.e. 24.1 kN for the tensile load-carrying capacity of the screws and 10% of this, i.e. 2.41 kN for the embedment capacity. Again, only the results for one screw per row are shown. The numbering of the rows is from the top edge to the bottom edge of each connector. The inclined screws in the secondary beam (the side where the load is applied) are all loaded in tension, but the tensile load-carrying capacity is not reached for any screw.

Interestingly, the load distribution is uneven, with rows 1 and 5 being the least loaded. On the main beam side, the loading follows the numbering of the rows. The tensile load-carrying capacity is reached for the top two rows. After the tests, tensile failure of screws in rows 2 and 3 was observed; see Figure 6.24a. For this connector, all inclined screws are loaded under tension. Again as with the other prototype, all screws reach the embedment strength perpendicular to the spring axis.

The evaluation of the contact springs shows no compressive failure perpendicular to the grain in the main beam underneath the connector plate. However, the bottom springs take up more load, leading again to a twisting of the connector. This behaviour was also observed during the tests; see Figure 6.24b. However, reinforcement screws were added for the tests to avoid failure perpendicular to the grain. Compressive forces parallel to the grain direction of less than half the compressive capacity occur at the end-grain end of the secondary beam, as shown in the specimens in Figure 6.24c. So again, the assumption of linear-elastic compressive behaviour parallel to the grain can be confirmed.

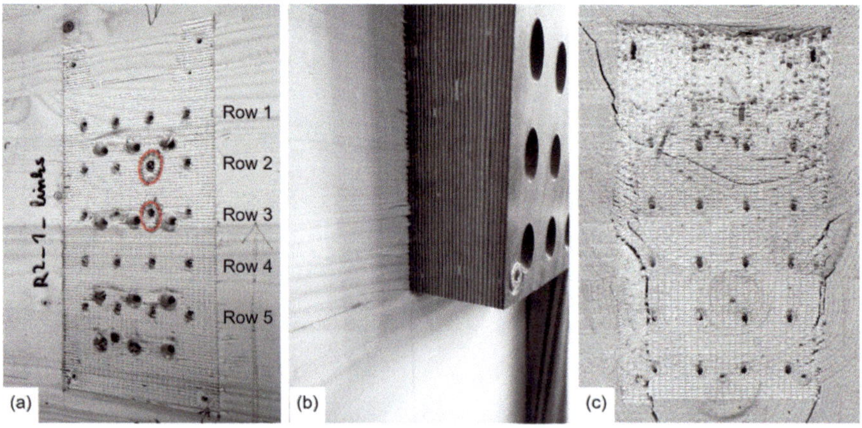

Figure 6.24: Connections after tests: (a) reaching of tensile capacity in rows 2 and 3; (b) no compressive failure perpendicular to the grain and (c) parallel to the grain.

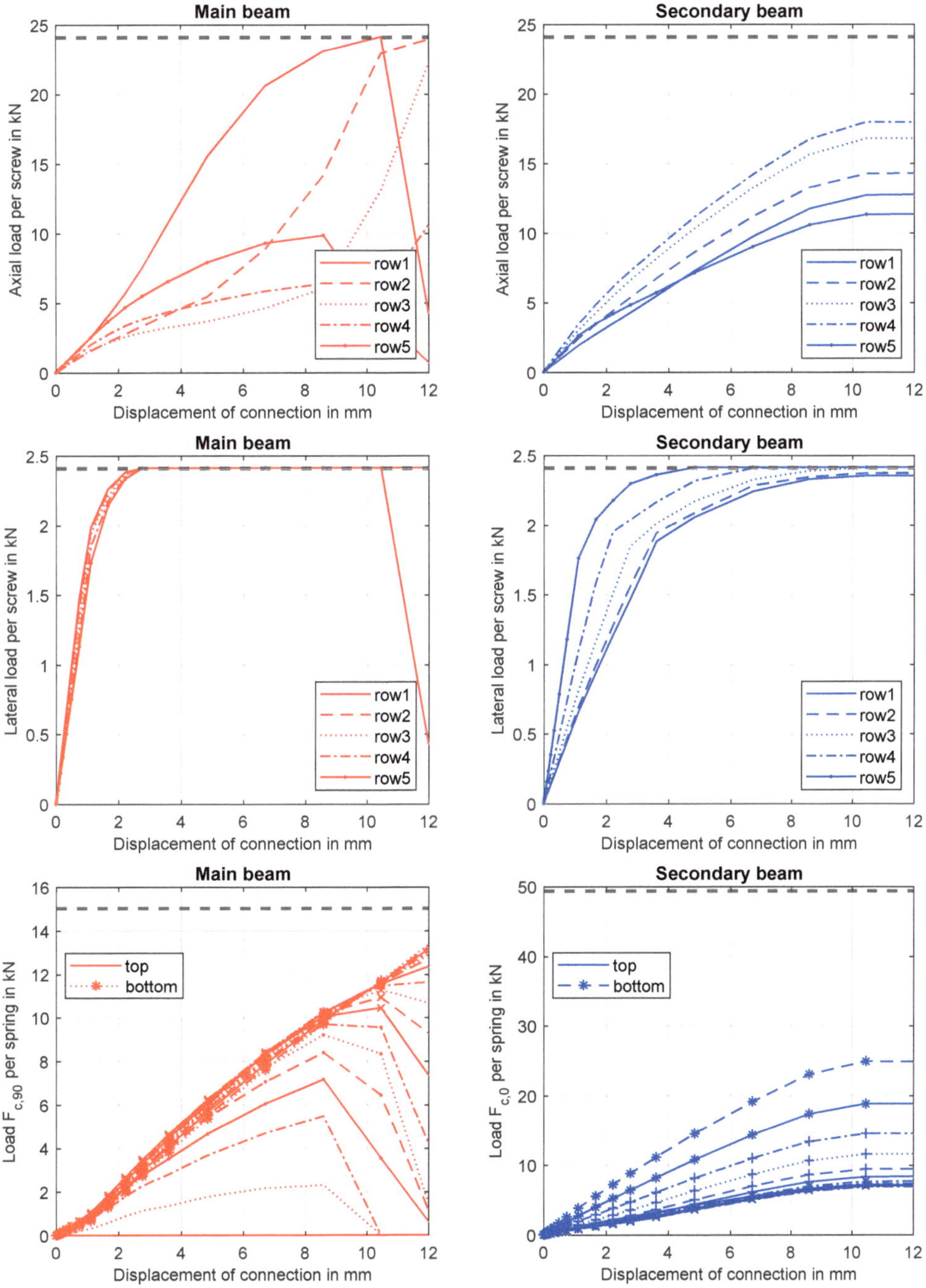

Figure 6.25: Spring loads of model with connector prototype v3.

6.3.5 Summary 2D

In summary, the 2D model predicts the load-carrying capacity very well, that is, for both modelling approaches, with and without rotation. In addition, the models can provide a reasonable estimation of the stiffness of the connection. Again, this goes for both modelling approaches. The input parameters of the different springs follow analytical equations. Those equations were taken from literature and followed the state of the art. Calibration of the springs was not needed and did, therefore, not occur. The validation of the springs with small-scale tests showed promising results. A good fit of the chosen input parameters was therefore assumed. The model works well if only vertical forces apply and no additional bending moments. The model cannot accurately simulate the rotation of the connector and its rotational stiffness and is, therefore, unsuitable for combined vertical loads and bending moments. A significant advantage is the low computational time and power needed for the model. Changes in the screws' properties are quickly implemented, and the results of the model are generated in a couple of minutes.

6.4 3D FE-Model

6.4.1 Structure of model

Since the 2D model only considers the connector as a simplified *beam element* and no statements can be made about the stresses in the connector itself, a 3D model was developed. This model considers the actual shape of the connector. Independently conducted material tests again form the basis for the input parameters of densified veneer wood [88] and fully threaded screws [57]. Tests with connectors with modified surfaces and inclined screws were modelled. The timber parts were modelled with *solid elements* (C3D20R – 20-node quadratic brick). The timber was modelled as orthotropic linear elastic with nine engineering constants. The DVW connector parts were modelled with *solid elements* (C3D20R) as orthotropic linear elastic with nine engineering constants. The modified surface of the connector was considered with the friction coefficients determined in Chapter 4. The input parameters are in Appendix A.4, Table A.10. The inclined screws were modelled with a steel core, surrounded by a "soft material". The bond between "soft material" and timber was defined by a *cohesive surface*. The input parameters are explained in Section 6.4.2, and the calibration is performed in Section 6.4.3. The additional mounting screws perpendicular to the shear plane were neglected.

Figure 6.26 shows the model and its boundary conditions. The load was applied by displacing the middle part by 20 mm. The corresponding displacement was recorded on the middle and side member, each halfway up the connector (highlighted with red dots in Figure 6.26). The symmetry of the test setup was utilised, and a symmetry plane was added. For the initial step, the screw tips were constrained but released once the load was applied. This constraint was necessary for the simulation to start.

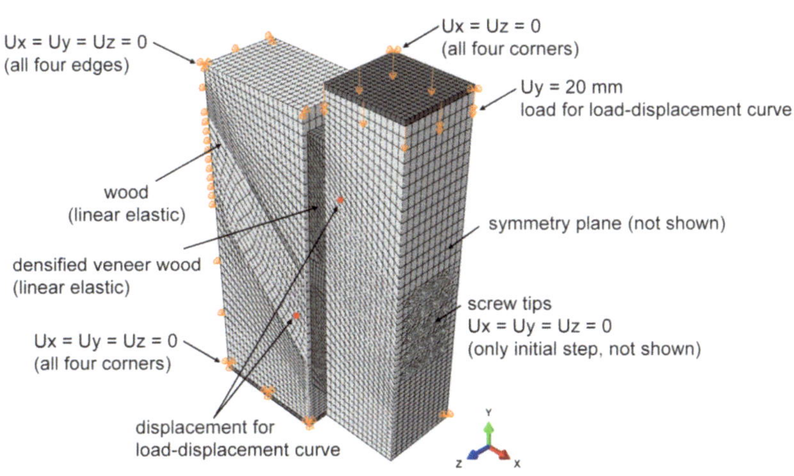

Figure 6.26: 3D-model with boundary conditions (credit to [94, 90]).

6.4.2 Modelling of axially loaded screws

The inclined fully threaded screws were modelled with a "soft material" and so-called *cohesive surfaces*, following [95, 96]. The screw core was modelled as a cylinder with a core diameter with *solid elements* (C3D20R) and isotropic steel properties. The screw thread and the wood matrix between the thread flanks were modelled together as a solid, the so-called "soft material", with reduced strength properties (taken from [95]). The screw core and the "soft material" are rigidly bonded together. The *cohesive surface* forms the bond between the "soft material" and the surrounding wood. Here, the properties of the bond, i.e. normal and shear stresses (K_{nn}, K_{ss} and K_{tt}), *damage initiation* (based on the maximum nominal stresses) and *damage evolution* (based on the fracture energy) are defined. The tests in [57] were considered to determine the input parameters. The values for K_{ss} and K_{tt} were calculated with Equation 6.14:

$$K_{\mathrm{ss}} = K_{\mathrm{tt}} = \frac{0.4 \cdot F_{\max}}{\delta_{04} \cdot (l_{\mathrm{ef}}/2) \cdot 2\pi \cdot r} \tag{6.14}$$

with δ_{04} displacement at $0.4 \cdot F_{max}$

ℓ_{ef} effective penetration length of the screw

r radius of the screw

6.4.3 Calibration of *cohesive surface*

The values for *damage initiation* and *damage evolution* are calibrated by modelling the withdrawal tests of screws with a diameter of 6 mm and an inclination angle of 45° from [57]. Figure 6.27 shows the calibration model and its boundary conditions. All materials were modelled with *solid elements* (C3D20R). Orthotropic properties were used for the wood, and isotropic properties for the steel parts. The load was applied by displacing the top of the screw by 5 mm. The displacement was recorded analogous to the experimental tests (highlighted in red in Figure 6.27). All input parameters are in Appendix A.4, Table A.15.

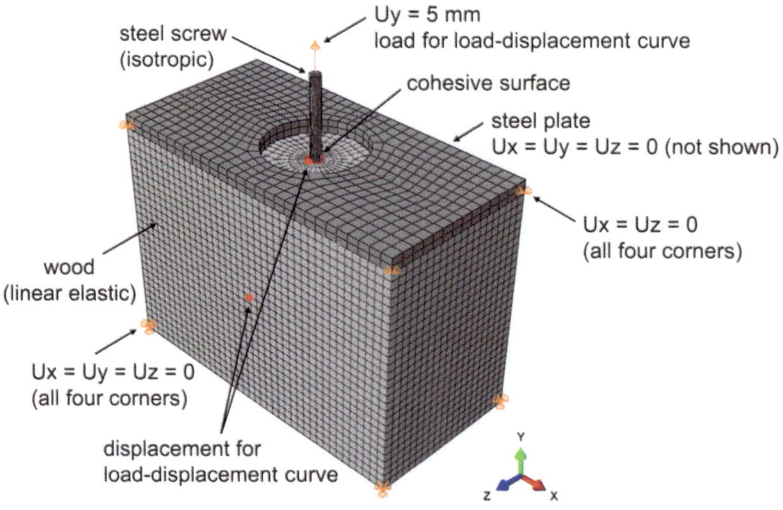

Figure 6.27: 3D model to calibrate the *cohesive surface* properties (credit to [94]).

The normal stress K_{nn} perpendicular to the surface was set to zero. The shear stresses K_{ss} and K_{tt} were calculated to 31.9 N/mm²/mm with Equation 6.14. For the first iteration step, the values for *damage initiation* (DI) and *damage evolution* (DE) were taken from [95]. However, these values relate to tests with screws with a nominal diameter of 13 mm and a core diameter of 9 mm. Figure 6.28 compares the results from the respective withdrawal tests with the numerical model. The model shows good agreement with the test results for values of the shear stress limit DI = 12 N/mm² and the fracture energy DE = 35 N/mm. The simulated load-carrying capacity of 10.1 kN is only 1% below the average load-carrying capacity from the tests. The deviation of the stiffness K_i is 25%, of K_s only 12%.

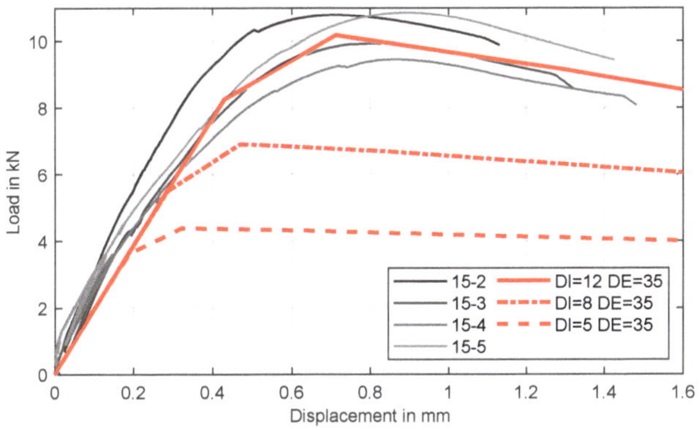

Figure 6.28: Comparison of test results from withdrawal tests with the numerical model (K_{ss} and K_{tt} were for all shown numerical calculations 31.9 N/mm²/mm).

6.4.4 Comparison with experimental results

The numerical results were compared to the experimental results from Chapter 5. Figure 6.29 shows the load-displacement curves in grey for the first tests with five screws 5x100 mm per connector plate and different surfaces. As can be seen in the diagrams, the beginning of the simulated curves corresponds well with the test results. However, the maximum load and displacement are overestimated. Although the model shows a clear drop in load during calibration, the maximum load of 50.8 kN is reached at the end of the calculation at a displacement of 20 mm. If the load is determined at the maximum displacement of the experimental tests, the conformity differs only by 8%, with a numerical load of 37.3 kN. The stiffness K_i is overestimated by approximately

25%, while the stiffness K_s is underestimated by approximately 5%. For the tests with the milled surface, the conformity between the test results and the model also looks good at first glance. However, the same problem arises: the calculations only stop when reaching the displacement limit of 20 mm. The load-carrying capacity of 66.5 kN is 25% higher than the load from the tests. Again, if the average displacement of the experimental tests is set as the limit, the numerical load is 55.9 kN and only 6% higher than the experiments. The deviations in stiffness values are similar. For the untreated surface K_i is overestimated by 23% while K_s fits quite well with a deviation of only 7%. For the milled surface, however, the deviations are significantly higher with 55% for K_i and 28% for K_s.

Both modelled load-displacement curves showed that the damage in the screws did not lead to the load to drop as expected. Therefore, the properties of the *cohesive surface* were fitted with the results from the push-out tests. The *damage initiation* or the shear stress limit *DI* was reduced to 5.0 N/mm^2. In Figure 6.29, the load-displacement curves of the fitted models are shown as dashed lines. The load drops after reaching a maximum and well before the displacement limit of 20 mm. The load-carrying capacity from the test and the model now differ by a maximum of only 5% and are both on the safe side. The deviations in the stiffness values remain the same as before.

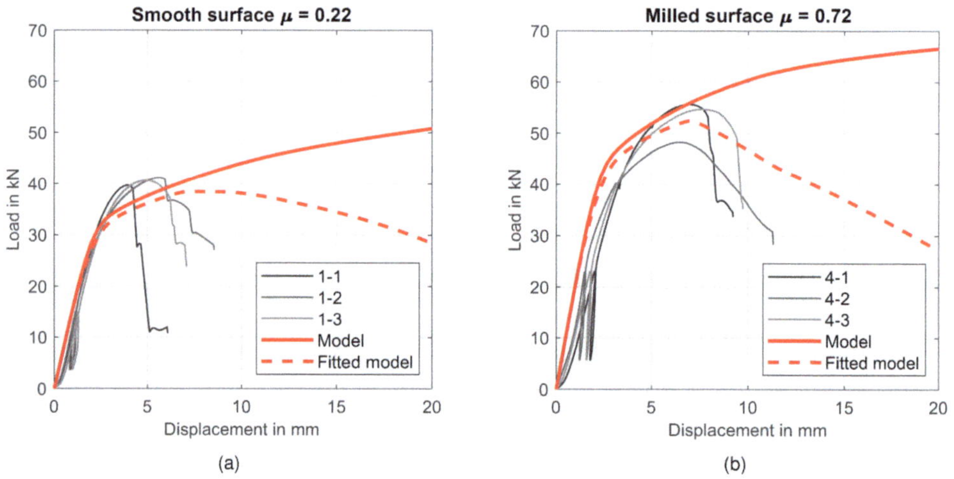

Figure 6.29: Comparison of experimental and numerical results for connectors with untreated surface (a) and milled surface (b).

With this fitted model, the push-out tests of the prototype connectors were simulated to gain information about the stress distribution in the connector plates and to use the model for a shape optimisation of the connector itself. Except for the measurements of the parts and the input parameters of the *cohesive surface*, no changes were made. The load-displacement curves of the tests with prototype v3 are in grey in Figure 6.30. Here, test 4 is the test where compressive failure perpendicular to the grain occurred, and test 5 is the test with CLT as side and middle members. The comparison to the model shows a perfect fit of the initial stiffness and the ultimate load. The simulated ultimate load was 472 kN and only 5% lower than in the experiments. The stiffness values are only 12–16% higher for K_i and K_s, respectively. As can be seen in the diagram, the model curve is almost congruent with the curve from the one test with CLT. Compared to the results of this single test, the ultimate load is 6% underestimated by the model, while both stiffness values are underestimated by only 10%.

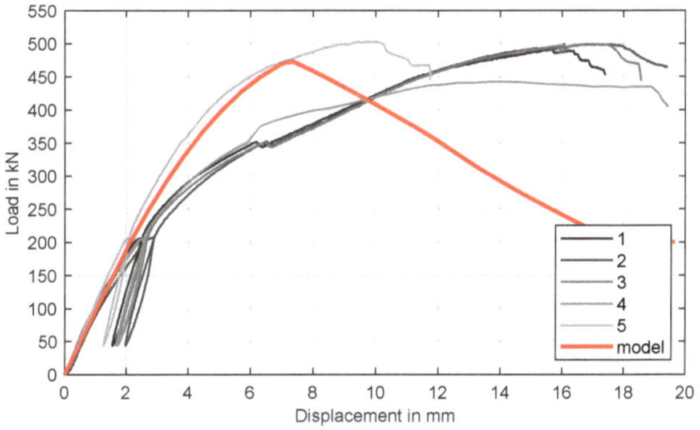

Figure 6.30: Comparison of experimental and numerical results for connectors with untreated surface (a) and milled surface (b).

One main objective of the 3D model was to gain insight into the stress distribution of the shear plane beneath the connector. Figure 6.31a shows the compressive stress in the shear plane of the timber side member. The scale ranges from $f_{c,90}$ = 0–4.3 N/mm^2. The contact pressure is distributed evenly across the shear plane. The indentation of the connector plate at its bottom can already be seen quite clearly. Here, the compressive strength was reached.

Furthermore, the evaluation of the contact pressure showed that already at small displacements, a contact pressure of 1.0 N/mm^2 is reached. Figure 6.31b shows the contact pressure in the shear plane plotted over the relative displacement of the connection. The data was evaluated at various points across the shear plane (highlighted with black dots in Figure 6.31a). The curves are cut off at the displacement at the ultimate load. At that point, they all are still well above 2.5 N/mm², the contact pressure used for most friction tests. These are valuable insights, as they show that the initially selected contact pressure for the friction tests occurs in that magnitude in the connection. Additionally, Figure 6.31b shows that a contact pressure up to 6.0 N/mm² occurs in the shear plane. Friction tests with surfaces intended for use in connections with inclined screws should, therefore, instead be conducted with a contact pressure greater than 1.0 N/mm² (although the coefficient of friction is independent of the contact pressure, which was also confirmed by the parameter study, nevertheless, friction tests should simulate the tribosystem[1] of interest).

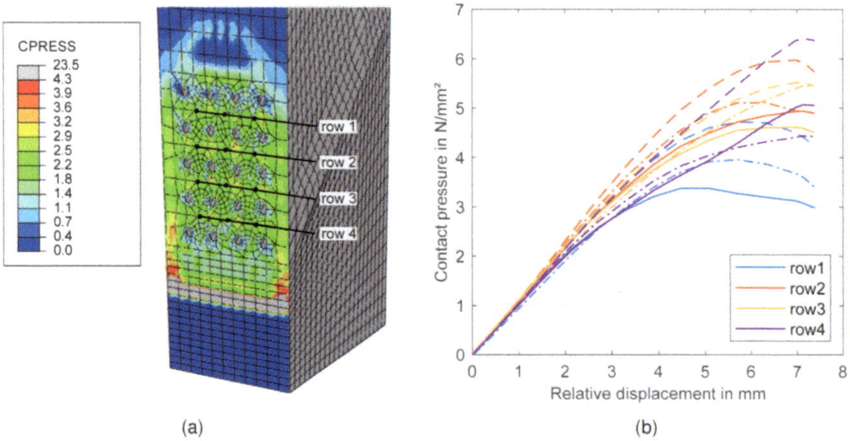

Figure 6.31: Contact pressure in the shear plane (a) and stress-displacement diagram (b).

Another objective of the 3D model was to gain insight into the stress distribution in the connector plate. With that information, a shape optimisation was intended. Figure 6.32a shows the normal stress distribution of the connector in its longitudinal direction, i.e. in the insertion direction, parallel to the load, and shear plane. The scale ranges from $f_{c,0}$ = 0–80.5 N/mm². The simulation shows that the compressive strength is reached in the insertion nozzle. This result was confirmed in the experimental tests, where a kinking band is visible; see Figure 6.32b. Additionally, a kinking band can be seen starting from the bottom row of screws to the bottom of the connector. Here, the simulation shows a compressive stress of about 80.5 N/mm², the mean compressive strength for DVW.

[1] A tribosystem is a system consisting of at least two contacting bodies and any environmental factors affecting their interaction.

The combination of the excellent agreement between the load-displacement curves with the test results and the internal stresses with the test results clearly shows that the model predicts the experimental tests well and can be used for further investigations. A disadvantage, however, is the highly increased modelling effort with the many individual parts. This increased effort also results in high computing time and resource requirements. Therefore, the model was not further used for a forming optimisation.

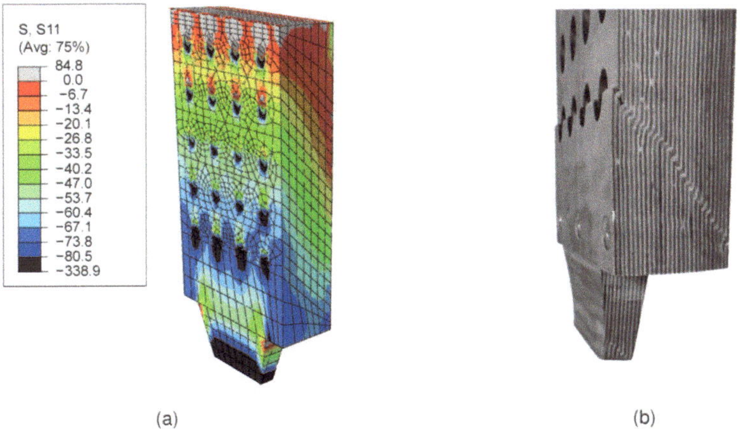

Figure 6.32: Compression stress of connector plate (a) and failed connector (b).

6.4.5 Summary 3D

In summary, the 3D model predicts the load-carrying capacity very well; however, only after a final fitting of the *cohesive surface*. In addition, the model can provide a reasonable estimation of the stiffness of the connection. The input parameters of the *cohesive surface* follow a calibration with small-scale test results. This procedure is also described in the literature and follows the state of the art. Unfortunately, further calibration of the *cohesive surface* was needed and was performed with the results of the first test series. The model predicts the internal stresses of the connection well, and the results can be confirmed with experimental tests. The model provides valuable insights but is impractical due to its extended modelling effort and computing time.

6.5 Conclusion

An analytical model for calculating the ultimate limit state of main beam to secondary beam connections with connectors made of DVW and modified surfaces was presented. The model aimed to predict the load-carrying capacity of a connection with inclined screws and increased friction in the shear plane. It was shown that the analytical model predicts the mean load-carrying capacity very well for different surfaces and friction coefficients, as well as for different lengths and numbers of fully threaded screws. Also, the model predicts the characteristic load-carrying capacity very well, with characteristic input parameters for the screws and timber. The Monte Carlo method could confirm the results for a range of friction coefficients, all in statistically determined limits. The simulation also clearly showed the influence of the coefficient of friction. The static coefficient of friction is suggested for calculating the load-carrying capacity (with the lowest ratio value of 0.75, mean ratio value of 1.09, and a R^2-value of 0.96). The evaluation of the model also showed that an increase in the friction coefficient due to surface modification cannot be transferred linearly to the calculation. This fact is illustrated in Table 6.4. Whereas the coefficient of friction can be more than doubled and tripled for various surface modifications, the ultimate load increases on average only by 30–40% (both comparisons regarding μ = 0.25). This discrepancy is because friction only amounts to a portion of the additional load-carrying capacity.

Table 6.4: Discrepancy in the increase of the friction coefficient (characteristic values) and the load-carrying capacity compared to the value of μ= 0.25 from Eurocode 5.

Surface modification	Increase of coefficient of friction		Increase of ultimate load[1)]	
	FG ⊥	EG	FG ⊥	EG
Milled pyramids 0.5 mm	156%	144%	31%	29%
Milled pyramids 1.0 mm	184%	136%	37%	27%
Milled pyramids 1.5 mm	216%	216%	43%	43%
Milled pyramids 2.0 mm	236%	-	50%	-
Milled circular	164%	144%	33%	29%
Embossed	136%	112%	27%	22%

FG = face grain and EG = end grain

[1)] for inclined screws with ε = 45°

The advantages of the analytical model are apparent. No complex modelling in a program is necessary, but the load-carrying capacities can be calculated directly (and even manually if necessary). The disadvantage is that the model does not provide any information about the deformation and stiffness of the connection. An analytical model for predicting the stiffness values should be evaluated for future use. In addition, a possible interaction of bending and tensile loading of the screws should be investigated.

A two-dimensional model with *beam* and *spring elements* was subsequently introduced. The main objective of the 2D model was to predict the stiffness of the connection. The springs can easily be calibrated and incorporated into the model. The results for both the load-carrying capacity and the stiffness values match well with the experimental test results. In addition, the individual spring loads coincide with the observations from the experiments. However, the model overestimates the rotation of the connection by a large extent. The degree of modelling is low, as all elements are only two-dimensional. The advantages, therefore, are the ability to predict the ultimate load and the deformation by keeping it simple and the computing time short. The disadvantage is that only failure of the springs is possible, therefore excluding by default any other failure mechanisms not covered by springs (such as tension perpendicular to the grain in the joist). Additionally, the model is only valid for exclusively vertical load and no additional bending moment/rotation. For future use, the rotational stiffness of the connection has to be further investigated to update the model accordingly.

Finally, a full-scale three-dimensional model was introduced. Again, the load-carrying capacity is predicted reasonably well. However, the resulting stiffness values differ from the experimental results to some extent. This model makes it possible to evaluate the stresses in both the connector and the timber parts. In addition, a shape optimisation can be performed with the 3D model. The advantage is the possibility of investigating the stresses in all the participating elements of the connection. However, this comes with disadvantages, such as the very long computing time and the need for high computing resources. This need is due to the complexity of the model, with every single screw being modelled on its own and the real shape of the connectors being used. Additionally, the model had to be fitted with the experimental results, as the calibration process with simple small-scale models did not work as expected (contrary to the beam and spring model).

However, all three investigated modelling approaches come to the same conclusion: it is not immediately and easily apparent which coefficient of friction is the correct one for the calculation. For the analytical model, using the static coefficient of friction μ_{stat}

resulted in a good fit of the model to the test results. For the 2D and 3D FE models, however, the mean coefficient of friction μ_{mean} was used and resulted in a good fit. Different friction coefficients should be considered for different limit states, e.g. the static friction coefficient for the serviceability limit state (SLS) and the kinetic friction coefficient for the ultimate limit state (ULS).

7 Long-term behaviour

7.1 Introduction

The increase in load-carrying capacity of connections with modified surfaces is based on increased friction in the shear plane due to the surface modification. Therefore, it is essential to have sufficient contact pressure in the shear plane throughout the connection's lifespan. Here, the question arises whether the contact pressure is still present after an extended time and, thus, friction is still activated. This behaviour should be the case since the fasteners are inserted at an angle to the fibre direction. This angle ensures that the fasteners are always loaded under tension and, therefore, exert a compressive force to the shear plane (see Equation 1.3 in Chapter 6).

Nevertheless, what happens when the timber shrinks and swells under climatic conditions? This question was investigated through long-term tests. Connections were exposed to different climatic conditions under constant load. This exposure was done with connections with inclined screws and with bolts inserted perpendicular to the shear plane, as already introduced in Chapter 5. In a further test series, assembled connections were conditioned to specific moisture contents of the wood before being tested for load-carrying capacity.

With the results of the tests under constant load, creep factors k_{def} could be determined. The factor k_{def} is used for the structural design of connections and reduces the connection's stiffness; see Equation 7.1:

$$K_{\mathrm{SLS,fin}} = \frac{K_{\mathrm{SLS}}}{(1 + k_{\mathrm{def,con}})} \tag{7.1}$$

with $K_{\mathrm{SLS,fin}}$ final mean value of slip modulus

K_{SLS} mean slip modulus of a connection

$k_{\mathrm{def,con}}$ creep factor of a connection

The creep factor k_{def} is calculated according to Equation 7.2 as the ratio of deformation increases to initial deformation. Literature uses periods of either one, three or ten minutes to determine the initial deformation. Analogous to [97], ten minutes was chosen to determine the initial slip.

$$k_{\text{def}} = \frac{(u - u_{\text{inst}})}{u_{\text{inst}}} \tag{7.2}$$

with u measured deformation of a connection

u_{inst} initial deformation of a connection

The creep factor is determined for 50 years. As experimental tests cover 50 years only in the rarest cases, it is necessary to extrapolate available test data. Therefore, different models are available, all differing in complexity and accuracy. Van de Kuilen [97] presented a creep model for timber-timber connections with dowel-type fasteners (Equation 7.3). The creep factor due to mechanically induced creep is based on a logarithmic function with two unknown parameters a and b. The validation with test data showed that a simple model with few input parameters is not necessarily less accurate than more complex models. Van de Kuilen & Dias [98] applied the model to timber-concrete connections, and an excellent fit to test results was confirmed. In contrast, the model presented by Pfefferle [99] is based on an exponential function including a damper to limit the exponential growth (Equation 7.4) and was initially developed for the creep of concrete. Reinhardt [100] adapted the model for the creep of wood.

- Logarithmic model [97]:

$$k_{\text{def}}(t) = a \cdot log(1 + b \cdot t) \tag{7.3}$$

- Exponential model [99]:

$$k_{\text{def}}(t) = a \cdot (1 - e^{-b \cdot \sqrt{t}}) \tag{7.4}$$

with t time in days

a parameter determined from test results

b parameter determined from test results

Each model must be calibrated with the available data to determine the parameters *a* and *b*. However, this is only possible for either the initial creep factor, covered in a short time (with available test data), or the final creep factor after 50 years. This dilemma can be seen clearly in Figure 7.1, where the two models mentioned above are compared with an additional two models from [101]. As Figure 7.1a shows, all four models predict the same creep factor for 570 days (the time covered by the tests herein). Nevertheless, as Figure 7.1b clearly shows, the behaviour after 50 years varies greatly. While both exponential models converge to a horizontal line, the power and logarithmic models predict further deformation. Once again, this emphasises the challenge to make qualified predictions about the behaviour of the connection in the future based on a timely limited database. The results should, therefore, be treated with caution. Hence, the logarithmic and the exponential models are compared to the test data. On the one hand, the logarithmic model was chosen as it was developed explicitly for timber-timber connections (see [97]). Conversely, the exponential model was chosen as it showed a good fit for timber in general (see [101]). The predicted creep factors should be used as lower and upper limits. The parameters *a* and *b* were determined for the time *t* in days.

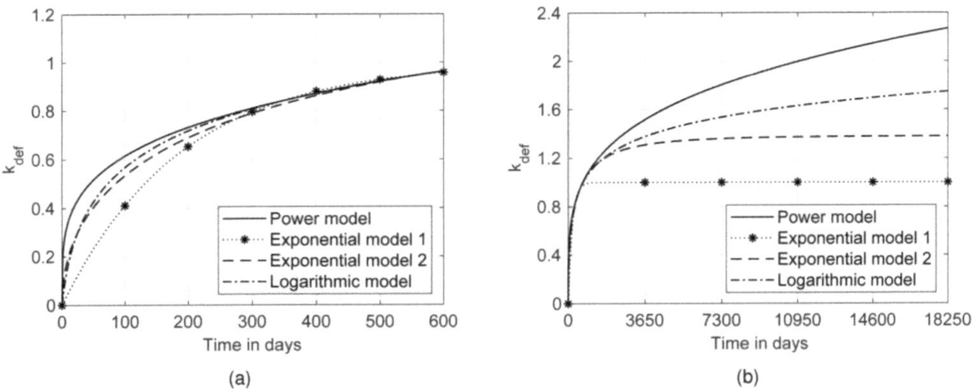

Figure 7.1: Different creep models and their behaviour over time: (a) period of 600 days; (b) period of 50 years.

7.2 Connections with inclined fasteners

7.2.1 Duration-of-load tests

7.2.1.1 Test specimens

The test specimens were manufactured similarly to the short-term tests. The connectors used were prototype v1 and v2, as introduced in Chapter 5. The connector plates were fastened to glulam GL 24h side and middle members with screws inclined at 45°. The used screws were of diameter d = 6 mm and length ℓ = 200 mm. The glulam was stored at a normal climate of 20°C and 65% r.h. before the tests. The density was determined in advance and was between 433–469 kg/m^3 for the middle members and 416–485 kg/m^3 for the side members. The side and middle members were matched with the density to be as close to each other as possible. The side members were significantly longer than in the short-term tests, so there were two connector plates on each side member, one at the top and one at the bottom. In total, four connections were tested with each specimen. One specimen in each service class contained solely four connectors v2, and the other specimen in each service class contained two connectors v2 at the top and two connectors v1 at the bottom.

7.2.1.2 Test programme and setup

The test programme consisted of duration-of-load tests and, subsequently, tests to establish the residual load-carrying capacity, analogous to the short-term tests. The test setup was based on the short-term tests and can be seen in Figure 7.2. The load was applied through large compression springs and threaded rods. The threaded rods had a metric thread M20 and strength class 10.9. The two middle members were loaded simultaneously and as evenly as possible via hollow steel profiles and large washers. This way, two push-out tests with four connectors could be tested in one system. The compression springs each had a spring force of 50 kN. Before the tests, each compression spring was loaded twice to 50 kN, with a constant loading speed. The springs were then divided into pairs according to their load-displacement behaviour. The load levels were chosen at $\sim$30% of the load-carrying capacity established in the short-term tests (33% for connector v1 and 28% for connector v2). The load of 50 kN per connector corresponds to a characteristic load level of the connectors of 37% for connector v1 and 33% for connector v2 (see Table 7.1). The load level was restricted to approximately 30% as only compression springs with a spring load of max 50 kN were available.

(a) (b) (c)

Figure 7.2: Loading of the specimens (a) and setup of tests with inclined screws in SC1 (b) and SC2 (c).

The test specimens were loaded with a universal testing machine up to 100 kN (i.e. 50 kN in each spring), and then the nuts of the threaded rods were tightened, keeping the springs compressed at 50 kN.

Table 7.1: Overview of long-term tests with connectors with milled surface.

Connector	Fasteners [mm]	$F_{max,mean}$ [kN]	$F_{max,char}$ [kN]	F_{spring} [kN]	load level (mean)	load level (char.)
Connector v1	10x 6x200	150	135[1)]	50	33%	37%
Connector v2	12x 6x200	180	154[2)]	50	28%	33%

[1)] 5th-percentile according to EN 14358 with $k_s(n) = 2.13$

[2)] 5th-percentile according to EN 14358 with $k_s(n) = 2.71$

The relative displacement between the side and middle member was measured with a digital depth gauge. The displacement was measured on the first day of the tests in short intervals, i.e. 10, 15, 30, 45, 60, 90 minutes. In the first week, measurements were taken twice a day. After the first week, measurements were taken twice a week, and after four weeks, measurements were taken once every week. After around five months, measurements were taken twice a month.

The test specimens were returned to the laboratory after ~330 days to check the load in the springs and reload them, if necessary. The recorded loads were between 90% and 95% of the target load. With a universal testing machine, each specimen was loaded back to 100 kN.

After ~570 days, the specimens were unloaded, again with a universal testing machine. The spring loads were no less than 93% of the target load. After unloading and removing the springs, the specimens were returned to their storage places to recover in their respective climatic environment. During recovery, the displacement was measured every 24 hours.

Service class 1 (SC1) is defined as indoors and heated, with an average temperature of 20°C and a relative humidity that exceeds 65% only for a few weeks each year. This condition leads to an average moisture content of the timber of < 12%. Service class 2 (SC2) is defined as roofed but open structures with a relative humidity that exceeds 85% only for a few weeks each year. This condition leads to an average moisture content of the timber of < 20%. The actual moisture content of the specimens was not measured during the tests. After the tests, the moisture content and the density were determined. The moisture content determined for the specimens in SC1 was, on average, 10.9%, while the moisture content for the specimens in SC2 was, on average, 15.4%.

Table 7.2 shows both connectors' ultimate load and deformation resulting from the short-term tests. Additionally, the displacement at a load level of 50 kN is given.

Table 7.2: Displacement of short-term tests at respective load levels.

	Connector v1						Connector v2					
	F_{max} [kN]	v_{max} [mm]	$F_{33\%}$ [kN]	$v_{33\%}$ [mm]	ρ [kg/m^3]	u [%]	F_{max} [kN]	v_{max} [mm]	$F_{28\%}$ [kN]	$v_{28\%}$ [mm]	ρ [kg/m^3]	u [%]
MEAN	151	13.1	50.1	2.32	451	11.2	173	8.95	49.9	1.21	456	11.3
SD	5.1	1.81	1.70	0.50	15	0.4	8.5	1.28	2.85	0.24	38	0.5
COV	3%	14%	3%	22%	3%	3%	5%	14%	6%	20%	8%	4%

7.2.1.3 Results and discussion

The tests started on 26. April 2022 and were ended on 13. November 2023, resulting in a total duration of 566 days. During that time, no failure of the specimens occurred.

Service class 1

The location where the test specimens were stored was indoors but not air-conditioned. The temperature and relative humidity were recorded over the whole time. The curves are shown in Figure 7.3. The temperature ranged between 14–29°C and the relative humidity between 28–68% (extreme values from the data set for both parameters). The relative humidity, therefore, did not exceed a value of 65% for several weeks per year.

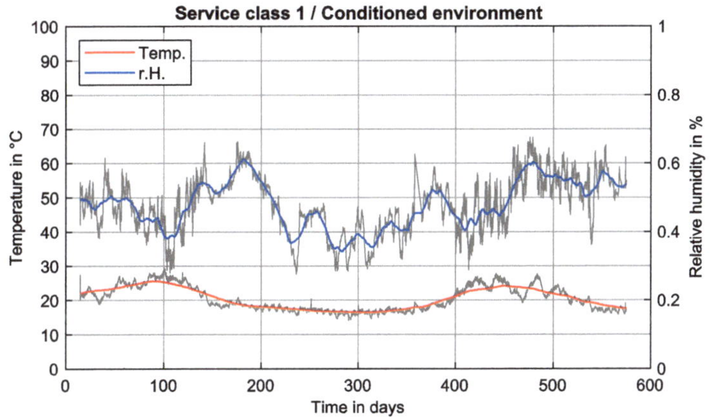

Figure 7.3: Temperature and relative humidity during the testing period in the conditioned environment.

Figure 7.4 shows the connectors' displacements in grey. In red is the average displacement for each connector type. The relative displacement of connector v1 increases sharply in the first few days and then slower over time. The displacement seems to converge at a displacement of 3.5 mm. The displacement of connector v2 is smaller, as there are more screws per connector. The increase in displacement is also more significant at the beginning and then slowly rises to around 3.0 mm. There is no direct correlation with the environmental conditions, i.e. no sudden increase or decrease of the deformation connected to an increase/decrease of the relative humidity or temperature. The jumps in the curve could also be due to the measurement of the deformation, as the measurement with the digital depth gauge and the angle brackets was quite error-prone. That is, mostly depending on the person doing the measurements and where exactly the depth gauge was placed. The drop in displacement in both diagrams is due to the unloading of the specimens after 566 days. The displacement in the shear plane drops to less than 2 mm for the connector v1 and less than 1.5 mm for the connector v2.

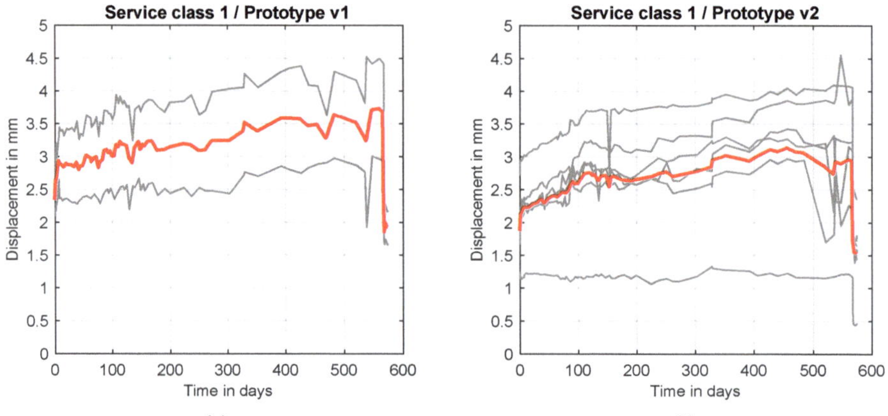

Figure 7.4: Displacement of connectors in long-term tests: (a) connector v1; (b) connector v2.

The initial displacement after 10 minutes $v_{10\text{min}}$ is given in Table 7.3. Connector v1 has an average displacement of 2.63 mm, while the displacement of the connector v2 is 1.93 mm after 10 minutes. This result corresponds quite well with the measured displacements of the short-term tests (2.32 mm and 1.21 mm). With this initial displacement, the creep factor k_{def} was calculated. The average moisture content of 11% confirms the environmental conditions' categorisation as SC1.

Table 7.3: Displacement of long-term tests in SC1 after 10 minutes.

	Connector v1			Connector v2		
	$v_{10\text{min}}$ in mm	ρ in kg/m^3	u in %	$v_{10\text{min}}$ in mm	ρ in kg/m^3	u in %
MEAN	2.63	440	10.9	1.93	439	11.0
SD	0.61	18	0.1	0.48	13	0.2
COV	23%	4%	1%	25%	3%	2%

Figure 7.5a plots the creep factor over time. Hereby, the creep factor corresponds to the average of all measured displacements of all connections, i.e. two connections

for connector v1 and six connections for connector v2. In red is the mean curve for connector v1, and in blue is the mean of connector v2. The expected creep factor for the test period is approx. 0.34 for connector v1 and 0.55 for connector v2. The exponential behaviour at the beginning of the tests is visible, as is the convergence to a horizontal line at the end of the tests. In addition, the simulated creep factors of the 566 test days for both connectors are given in Figure 7.5a with dashed lines. The two different models fit quite well with the test results.

The values for the input parameters *a* and *b* are given in Table 7.4. Also given in Table 7.4 are the R^2-values that rate the fit of each model. The R^2-values are similar, ranging from 0.81 to 0.86. Figure 7.5b shows the extrapolated curves for 50 years. Accordingly, the creep factor ranges between 0.68 and 0.93 for connector v1 and between 0.72 and 1.0 for connector v2.

Table 7.4: Input parameters for different creep models (SC1).

Model	*a*	*b*	R^2	Connector
Logarithmic model	0.402	0.011	0.83	v1
	0.300	0.118	0.85	v2
Exponential model	0.702	0.027	0.86	v1
	0.725	0.059	0.81	v2

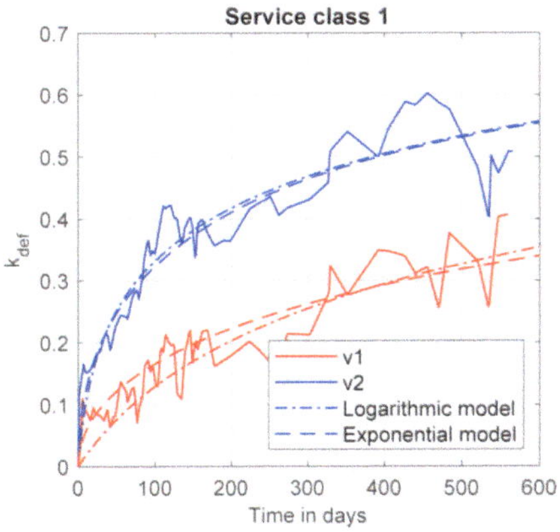

(a)

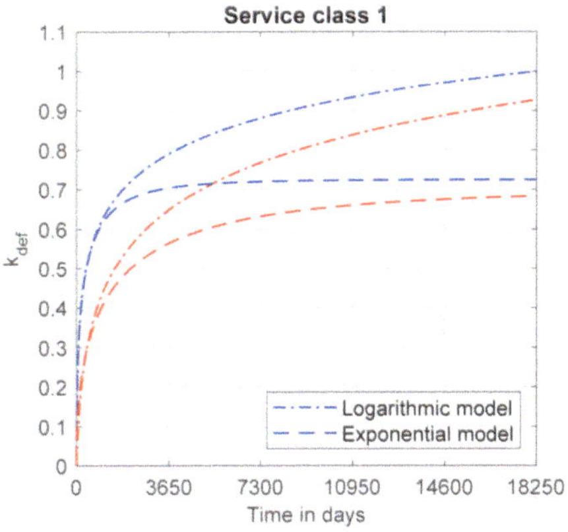

(b)

Figure 7.5: Creep factor k_{def} for 570 days (a) and extrapolated for 50 years (b) for connector v1 (red) and v2 (blue).

Service class 2

The test specimens were stored for almost 600 days outdoors in a covered shed. The temperature and relative humidity were recorded over the whole time. The curves are shown in Figure 7.6. The temperature ranged between -4° and +39°C and the relative humidity between 15–95% (extreme values from the data set for both parameters). The relative humidity, therefore, did exceed a value of 65%, but not 85%, for several weeks per year. Due to an error of the data logger, no data was saved for ~100 days in the first year.

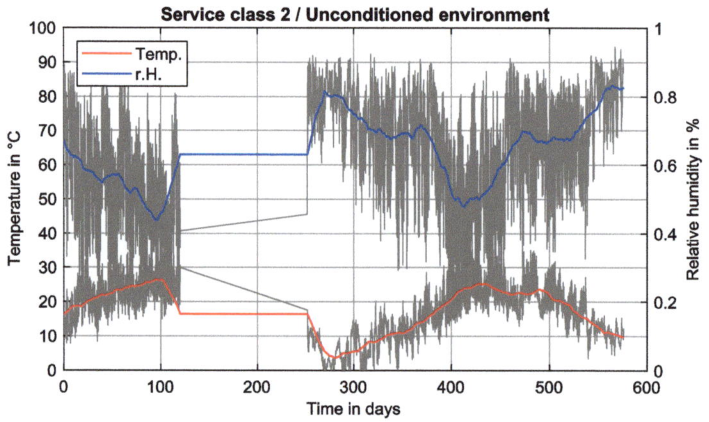

Figure 7.6: Temperature and relative humidity during the testing period in the unconditioned environment.

In Figure 7.7 the displacement over time is given for all connectors in grey. The average displacement of each connector type is given in red. The relative displacement of connector v1 increases sharply at the beginning and constantly at a much lower rate. The displacement converges to approximately 5.5 mm. Again, the displacement of connector v2 is smaller due to the greater number of screws. After the initial displacement, the displacement increases moderately over time. The displacement seems to converge at approximately 4.0 mm. No direct correlation between the environmental conditions and the displacement can be made. The drop in the deformation of connector v1 at about 400 days coincides with a drop in the relative humidity simultaneously. However, for the connector v2, no such drop of deformation was recorded. That implies that the drop is instead a measuring error.

After about 330 days, the springs were reloaded to their initial load level. Afterwards, a significant increase in deformation was observed in all connections. Especially one connection deformed more than the others, from about 2 mm to 5 mm within 100 days.

The same connection was exposed to the weather because of a hole in the shed's roof, and rainwater accumulated on top of the middle member of that connection. It cannot be excluded that water also entered the connection. This exposure explains the extraordinary rise in the average curve during this period. The drop in deformation at the end is again due to the unloading of the specimens. A remaining deformation of approx. 3 mm and 2 mm was recorded for the connectors v1 and v2, respectively.

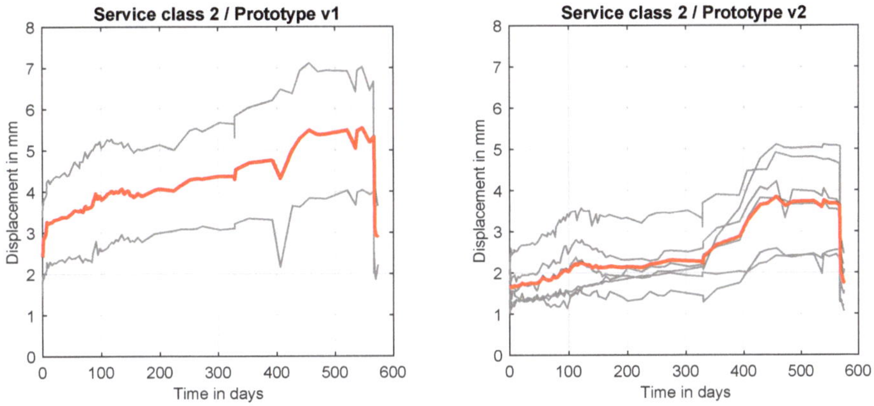

(a) (b)

Figure 7.7: Displacement of connectors in long-term tests: (a) connector v1; (b) connector v2.

The initial displacement after 10 minutes v_{10min} is given in Table 7.5. The connector v1 had an average displacement of 2.65 mm, while the displacement of the connector v2 was 1.66 mm. These results correspond well with the measured displacements of the short-term tests (2.32 mm and 1.21 mm).

Table 7.5: Displacement of long-term tests in SC2 after 10 minutes.

	Connector v1			Connector v2		
	v_{10min} in mm	ρ in kg/m^3	u in %	v_{10min} in mm	ρ in kg/m^3	u in %
MEAN	2.65	452	16.3	1.66	469	15.2
SD	1.13	24	1.7	0.39	23	0.5
COV	43%	5%	10%	23%	5%	3%

Figure 7.8 shows the creep factor over time. The calculated creep factors of the 566 test days are shown in Figure 7.8a. The creep factor was determined with the average

displacement of all measured connections, i.e. two connections for connector v1 and six connections for connector v2. For both connectors, the creep factor varies between 0.9 and 1.1 (v1 = red line and v2 = blue line). After 330 days a spike in the measured displacement (Figure 7.7b), and, therefore, in the creep factor (Figure 7.8a) was observed. This happened right after the specimens were reloaded to their designed test load of 50 kN in each spring. Following the reloading, the displacement significantly increased over time. When trying to fit the models to the evaluated creep factor, the R^2-values show a reasonable agreement for connector v1 with 0.81–0.87, but a poor agreement for connector v2 with 0.53–0.69.

The values for the input parameters *a* and *b* are given in Table 7.6, as are the R^2-values. Figure 7.8b shows the extrapolated curves for 50 years. According to that, the creep factor ranges between 1.4–2.8 for both connectors.

Table 7.6: Different models for creep and their input parameters (SC2) for 570 days.

Model	*a*	*b*	R^2	Connector
Logarithmic model	1.17	0.011	0.87	v1
	1.20	0.011	0.69	v2
Exponential model	1.45	0.043	0.81	v1
	1.39	0.046	0.53	v2

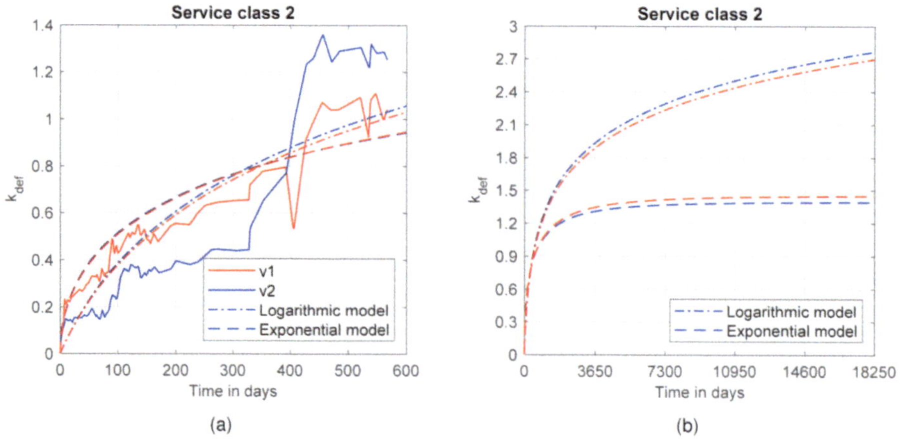

Figure 7.8: Creep factor k_{def} for 570 days (a) and extrapolated for 50 years (b) for connector v1 and v2 with original data.

Figure 7.9a shows the creep factor evaluated only for a time range of the first 330 days. Now, the fit of the models is much better, and, therefore, the creep factors for a period of 600 days reduce to around 0.8 for connector v1 and 0.6 for connector v2. The values for the input parameters *a* and *b* are given in Table 7.7, as are the R^2-values. Here, the good fit of the models can be confirmed, with R^2-values of 0.99. Figure 7.9b shows the extrapolated curves for 50 years. According to that, the creep factor now ranges between 1.2–1.4 and 1.2–1.8 for connectors v1 and connector v2, respectively.

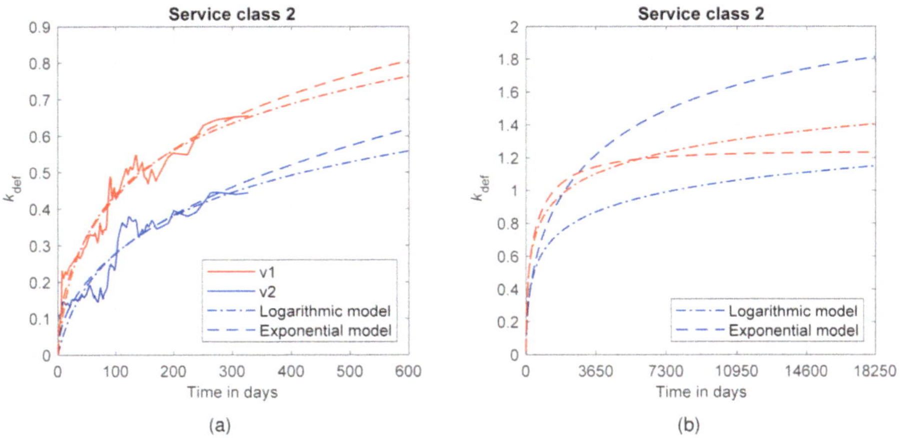

Figure 7.9: Creep factor k_{def} for 330 days (a) and extrapolated for 50 years (b) for connector v1 and v2 with adjusted data.

Table 7.7: Different models for creep and their input parameters (SC2) for 330 days.

Model	*a*	*b*	R^2	Connector
Logarithmic model	0.44	0.092	0.98	v1
	0.40	0.039	0.99	v2
Exponential model	1.24	0.043	0.99	v1
	2.13	0.014	0.99	v2

Skewing of the test setup

During the tests, a skewing of the middle members of the test specimens was noticed. This skewing can be seen in Figure 7.10. Visible gaps open between the connector plates and the timber parts. The twisting is probably due to several reasons. On the one hand, it may have happened that the load introduction into the test specimens via the steel profiles was not precisely centred. Thus, a slightly eccentric load introduction loaded the test specimens one-sided. Also, the test specimens were very small. On the other hand, it is also possible that the springs were not evenly loaded. Another possible explanation for the twisting of the middle parts is that there were differences in stiffness on the right and left per pair of connectors. In addition, two different types of connectors with different stiffnesses were tested on one test specimen.

When the springs were reloaded, explicit care was taken to ensure that the load application was centred, but the skewing continued to increase afterwards. The skewing also influenced the measurement of the relative displacement to a certain extent, so the values given here should be considered against this background. In future experiments, the test setup should be modified so that the springs are not pushing the specimens together but rather pulling them apart, thus stabilising the system automatically through tensile forces.

After unloading and during the recovery phase of the connections, all middle members rotated back to their initial position. At the beginning of each test of the residual load-carrying capacity, no visible gaps between the connectors and the timber were observed. The test setup should be alternated for future investigations so that the springs enforce tensile loads on the specimens. In that way, the system stabilises itself.

Figure 7.10: Skewing of middle members of test setup under compression.

7.2.2 Residual load-carrying capacity

The test specimens were unloaded after 566 days. The unloading was done by placing them in a universal testing machine and slowly unloading the springs at a constant rate. The specimens were left to recover, and the displacement was regularly measured. The recovery occurred in the same climatic conditions, i.e. SC1 and SC2. The specimens were left to recover for 30 days. This recovery phase was mainly due to logistical reasons in the laboratory. Figure 7.11a shows the theoretical displacement curve for wood under a constant load. After unloading, the deformation decreases until a small share due to plastic deformation is left. As can be seen in Figure 7.11b, the displacement u_{plast} did no longer change shortly after unloading. The plastic deformation was between 2.0 and 2.5 mm for connector v1 and about 1.5 mm for connector v2. Just before the tests, the specimens were cut in half to receive the identical push-out specimens as for the short-term tests. The tests to determine the residual load-carrying capacity were performed analogous to Chapter 5, Section 5.3 and EN 26891 with unloading loop. During the tests, the load and the displacement of the middle and side members were recorded constantly. The displacement was measured with a DIC system and evaluated at the exact locations as before in the short-term tests. After the tests, the moisture content and the density were determined individually for each connection.

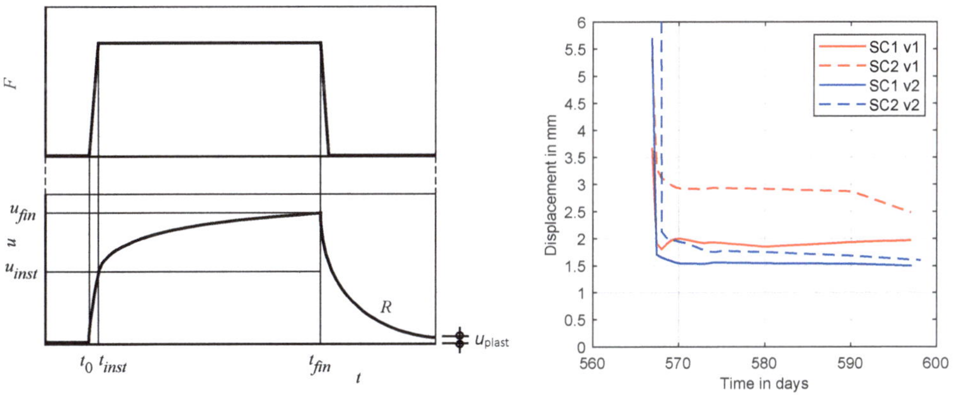

(a) (b)

Figure 7.11: (a) Load and displacement over time [102] and (b) residual displacement during recovery phase.

Service class 1

The single test with connector v1 reached a mean load of 143 kN, which is only 5% lower than the short-term tests and lies within the standard deviation of these tests. The load-displacement curve (red line) is given in Figure 7.12a. The figure also shows the load-displacement curve of Series 6 of the short-term tests (black line). The curves are almost congruent, however, the stiffness seems to be lower at the beginning. The connection first deforms before taking up load. This is where the gradient of the long-term curve increases more than for the short-term tests, see Figure 7.12b. With $K_s = 28.2 \pm 1.8$ kN/mm, the stiffness value is 5% higher than the stiffness of the short-term tests. As before, the stiffness was evaluated strictly in the range of 10–40% of the ultimate load. At approximately the same displacement of the short-term tests, the load drops, and the connection fails due to the compressive failure of the connector.

Table 7.8: Test results of residual load tests in SC1.

	Connector v1 (n = 1)				Connector v2 (n = 3)			
	F_{max} [kN]	v_{max} [mm]	K_s [kN/mm]	u [%]	F_{max} [kN]	v_{max} [mm]	K_s [kN/mm]	u [%]
MEAN	143	10.7	28.2	10.9	145	11.1	35.7	11.0
SD	-	0.3	1.8	0.1	4.1	1.22	3.0	0.2
COV	-	3%	7%	1%	3%	11%	8%	2%

As initially expected, the strength and stiffness are the same for the short-term and the tests in SC1. The variation by ±5% is not significant because only one test was carried out. The results show no influence of the climatic condition of SC1 on the behaviour of the connection with the prototype connector v1.

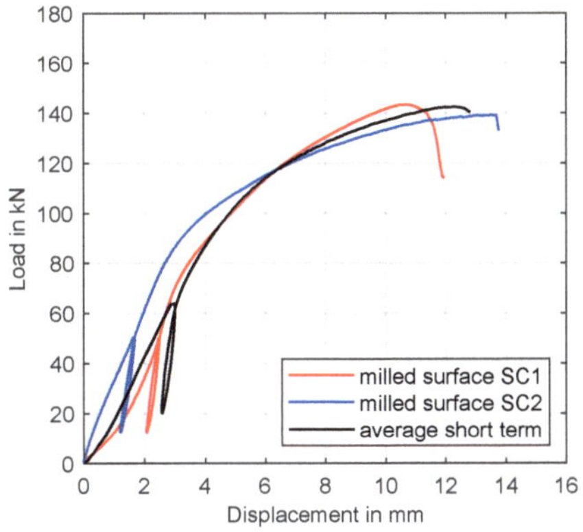

(a)

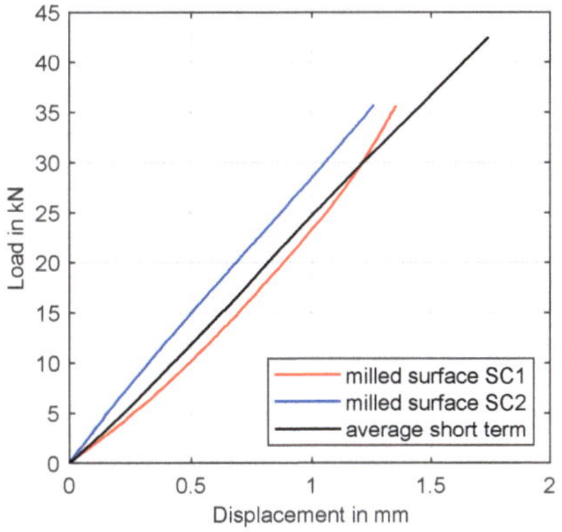

(b)

Figure 7.12: Residual strength (a) and stiffness (b) for tests in SC1 and SC2 with connector v1.

The connections with connector v2 exposed to SC1 reached a mean load of F_{max} = 145 ±4 kN. That is only 84% of the mean load of the short-term tests. In Figure 7.13a, both the short- and long-term test curves are given. Statistically, there is a significant difference between the results of the short-term tests and the long-term tests (p = 0.019 and F = 6.36). The Tukey's HSD Test showed that the difference is between the short-term and SC1 tests. The connection is initially softer than the short-term tests, and the curve flattens relatively soon. At a slightly larger maximum displacement, tensile failure occurs in the inclined screws. The significantly flatter behaviour at the beginning of the test can be seen in Figure 7.13b. A possible explanation could be that small shrinking deformation of the timber, and the elastic re-deformation after the springs were relieved, had to be overcome first during loading. However, at 35.7±3 kN/mm, the stiffness between 10 and 40% of the maximum load is approx. 7% greater than in the short-term tests. All evaluated test results and the measured moisture content at the time of the tests are given in Table 7.8.

The lower load-carrying capacity might be due to different reasons. On the one hand, only three tests were performed and evaluated. On the other hand, timber's mechanical properties reduce over time [103]. In the design of timber structures and connections, this is considered through the modification factor k_{mod}. Therefore, a reduced load-carrying capacity seems reasonable. Finally, a possible explanation could be that the timber swelled and shrank throughout the test period and the changing climatic conditions. The wood matrix might have withdrawn slightly from the modified surface structure (in this case, the pyramid pattern). During the loading of the tests, the connectors first had to be pressed into the softwood again before friction was fully activated (this could also explains the recorded slip at the beginning of the tests, which can be seen in Figure 7.13). After the subsequent activation of friction, a greater displacement was required to achieve the same ultimate load. However, the screws may have already reached their deflection limit, potentially resulting in failure before the maximum possible load was achieved. This theory goes hand in hand with the investigations on the interaction between bending load and normal load on screws [104].

The constant loading can explain the increase in stiffness during the test period, as the pyramid pattern was continuously pressed into the softwood. Once the shrinkage of the wood matrix had been overcome, the surface structure was fully interlocked with the softwood surface. In contrast to the short-term tests, where the surface structure caused deformations perpendicular to the shear plane and, thus, low stiffness values, there were now no deformations due to the impressing of the pyramids. This allowed the overall stiffness of the connection to increase.

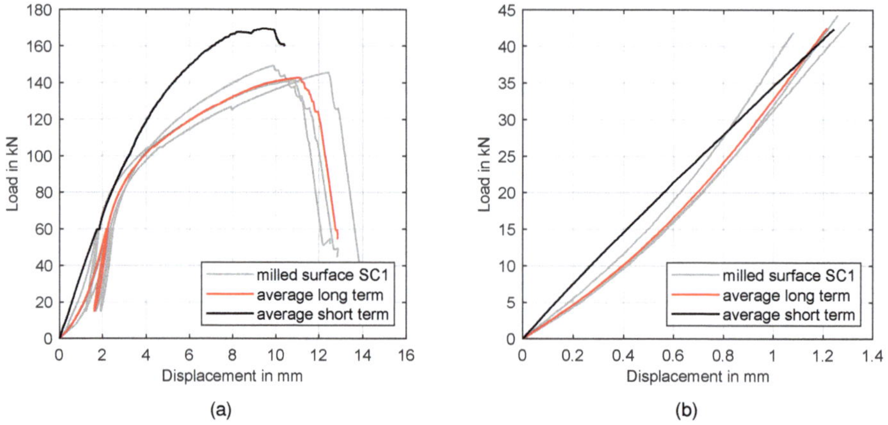

Figure 7.13: Residual load-carrying capacity (a) and stiffness (b) for tests in SC1 with connector v2.

Figure 7.14 shows the connection and the connector after the tests. In Figure 7.14a, the excellent imprint of the modified surface into the timber member can be seen at the bottom of the connector. This imprint shows, sufficient contact pressure can be created in the shear plane even after an extended time under constant load. Additionally, it shows that the modified surface was not affected by the changing environmental conditions and the constant load. Figure 7.14b shows a failed connector. During the short-term tests in two specimens the connector also failed due to compression parallel to grain. Significant damage to the connector due to the environment can be excluded.

(a) (b)

Figure 7.14: Imprint of milled surface in softwood (a) and failed connector (b).

Service class 2

The single test with connector v1 exposed to SC2 reached a mean load of F_{max} = 139 kN. That is 8% lower than the short-term tests. This result is also lower than the test in SC1, as expected. The load-displacement curve (blue line) is also shown in Figure 7.12a. The connection takes up the load right from the start of the test, and no slip occurs. This behaviour is confirmed by the stiffness of K_s = 28.2±1.4 kN/mm, which is 5% higher than the short-term tests, see Figure 7.12b. A possible explanation for this could be that due to the swelling of the timber, there was full contact between the connectors from the beginning of the test. Especially the modified surface with its pyramid pattern was fully pressed into the timber. All test results and the respective moisture content at the time of testing are given in Table 7.9.

Table 7.9: Test results of residual load tests in SC2.

	Connector v1 (n = 1)				Connector v2 (n = 3)			
	F_{max} [kN]	v_{max} [mm]	K_s [kN/mm]	u [%]	F_{max} [kN]	v_{max} [mm]	K_s [kN/mm]	u [%]
MEAN	139	13.6	28.2	16.3	154	11.8	36.4	15.2
SD	-	0.6	1.4	1.7	0.7	0.38	4.6	0.5
COV	-	4%	6%	10%	0%	3%	13%	3%

The tests with the connector v2 reached a mean load of 154 kN, which is only 90% of the mean load of the short-term tests. The Tukey's HSD Test for multiple comparison showed no significant difference between the short-term and the SC2 tests. The average load-displacement curve (red line) is given in Figure 7.15a. Also given is the short-term load-displacement curve (black line). At the beginning, the load-displacement curves are congruent before the curve of the residual load-carrying tests flattens significantly. The stiffness of 36.4±4.6 kN/mm is 10% higher than those of the short-term tests. The load plotted over the local displacement is given in Figure 7.15b. The explanations for the decrease of load-carrying capacity and the increase of stiffness are the same as for the tests in SC1.

Much more interesting is the comparison with the tests in SC1. The load-carrying capacity and the stiffness are higher than the results of the tests in SC1. This increase was not to be expected. The swelling of the wood can explain the increase in both cases. The wood matrix encloses the pyramid pattern better, leading to a better interlocking with the softwood. This can be seen in the load-displacement curves in Figure 7.15. The is no slip at the beginning as no deformation perpendicular to the shear plane occurs.

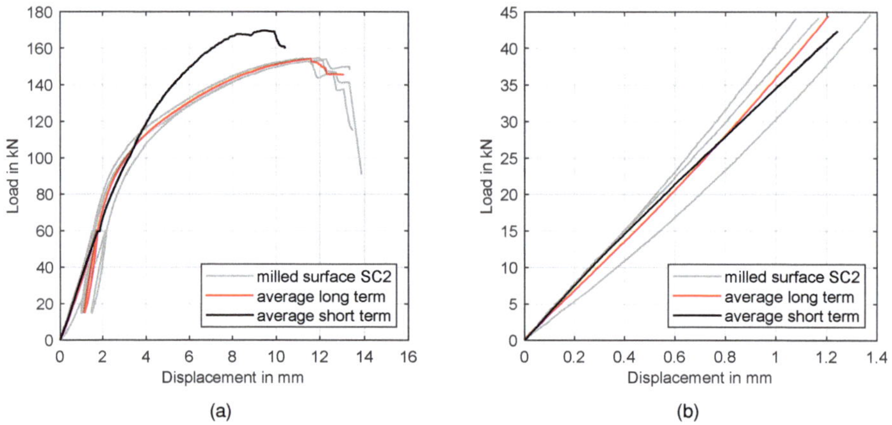

Figure 7.15: Residual load-carrying capacity (a) and stiffness (b) for tests in SC2 with connector v2.

Figure 7.16 shows the connection after the tests. In Figure 7.16a, a bottom plate of connector v2 can be seen with the excellent imprint of the modified surface into the softwood. A twisting of the connector can be recognised to some extent. The head of the lower assembly screw is protruding from the connector plate, and the lower edge of the connector is pressing into the wood. However, this could have happened after the connection failed. As shown in Figure 7.16b, the connector plate suddenly shifted upward on failure and compressed the wood fibres at the upper edge. At the lower edge, on the other hand, it seems like the right corner was not as firmly pressed into the wood as the other three corners. Figure 7.16c shows the modified surface of the same test. Again, the left side was pressed more into the wood as more fibres stuck to the pyramid pattern.

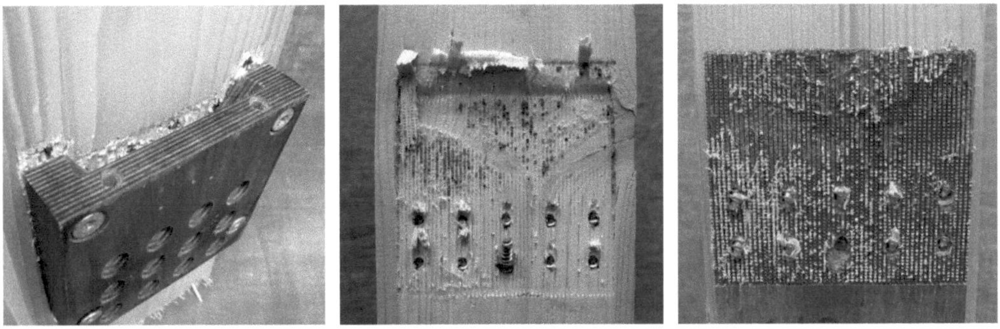

(a) (b) (c)

Figure 7.16: Connections after test: (a) impression of connector v2; (b) close-up of shear plane; (c) close-up of modified surface.

7.2.3 Influence of shrinking and swelling

Tests were carried out with conditioned specimens to investigate the influence on the connection of changing moisture content of the timber parts. Analogous to the short-term tests, the connectors v1 were fastened to the timber with fully threaded screws. The timber specimens were stored beforehand at 20°C and 65% r.H., resulting in a moisture content of 12–13%. After assembly, the specimens were stored in a controlled environment of 20°C and 95% r.H. for almost 12 months. At that point, the moisture content was, on average, 15.5%. Subsequently, the specimens were returned, first to an environment of 20°C and 65% r.H., and later to an environment of 20–30°C and 40% r.H., to speed up the drying process. This was supposed to resemble a possible installation situation at the construction site. The structure is erected with a conditioned moisture content of 12%, then the structure becomes wet during the further construction period, and the moisture content increases to e.g. 16%. After the construction is finished and the building is heated, the moisture content drops to 12% or even less. Figure 7.17b shows the actually measured moisture content of the three specimens over a period of 600 days. The specimens were tested for load-carrying capacity once the moisture content returned to 12–13%.

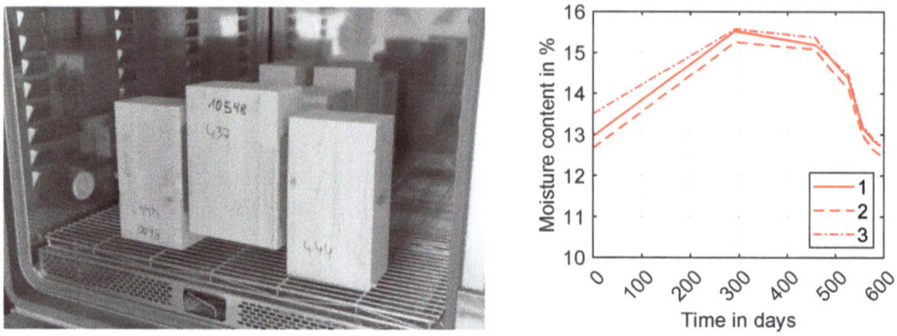

(a) (b)

Figure 7.17: Specimen in climatic chamber (a) and progression of moisture content (b).

The tests were performed analogously to the short-term tests and EN 26891 with an unloading loop. The load was measured constantly during the tests, as was the displacement of the side and middle members (measuring rate of 1 Hz). The displacement was measured with a DIC system with the same settings as described in Chapter 5.2.1. The results of the three tests (changing m.c.) are given in Table 7.10. Also given are the results of Series 6 of the short-term tests with connector v1 performed in Chapter 5.3.3 (constant m.c.).

Table 7.10: Test results of specimens with connector v1 exposed to changing moisture contents.

	Changing m.c. ($n = 3$)				Constant m.c. ($n = 9$)			
	F_{max} [kN]	v_{max} [mm]	K_s [kN/mm]	u [%]	F_{max} [kN]	v_{max} [mm]	K_s [kN/mm]	u [%]
MEAN	142	14.7	22.8	11.2	151	13.1	25.0	11.2
SD	5.6	2.97	0.9	0.2	5.1	1.89	4.0	0.4
COV	4%	20%	4%	1%	3%	14%	16%	3%

The load-displacement curves are given in Figure 7.18a. The overall shape of the curve is congruent with the results from the short-term tests. However, some slip was observed at the beginning of the test, analogous to the long-term tests of SC1. Also, the curve flattens around 100 kN, which was also observed for the other long-term tests, both in SC1 and SC2. The average load was 142±5.6 kN and only 6% lower than the short-term test load. The stiffness and, thus, the local displacement is given in Figure 7.18b. After the initial slip the slope of the curves is almost parallel, which shows in the stiffness values. Here, a mean stiffness of 22.8±0.9 kN/mm was determined, which is about 10% lower than the stiffness in the short-term tests.

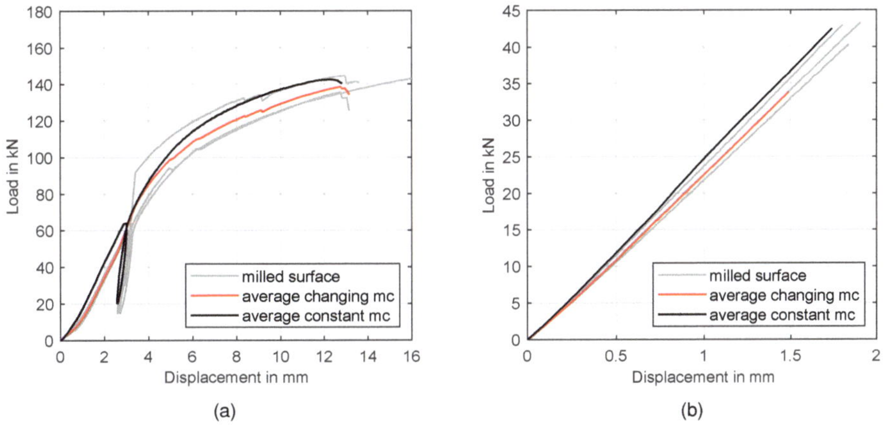

Figure 7.18: Residual load-carrying capacity (a) and stiffness (b) for tests in SC1 and SC2 for connector v1.

7.2.4 Summary

Duration-of-load tests were performed with connectors with modified surfaces. The objective of the tests was to investigate the long-term behaviour of connections with inclined screws and modified surfaces. The duration of the creep tests covered 570 days in service class 1 and service class 2, according to Eurocode 5. Creep factors were determined with an initial displacement after 10 minutes. No imminent influence of the temperature on the creep factor is visible, and no imminent influence of the relative humidity on the creep factor can be seen. That is, the behaviour of the creep factor curve is not directly linked to the behaviour of the temperature or moisture content curve. The creep factors were extrapolated with two models (logarithmic or exponential) for 50 years. The results (mean values of both models) are given in Table 7.11.

Subsequently, push-out tests were carried out to determine the residual load-carrying capacity of the connections with increased friction in the shear plane. The results showed a decrease in the ultimate load of 5–16% for service class 1 and a decrease of 8–10% for service class 2. However, the stiffness in the range of 10–40% of the ultimate load was increased in all tests, i.e. 5–7% for the tests in SC1 and 5–10% for the tests in SC2. The failure modes were analogous to the short-term tests.

Finally, push-out tests were conducted with specimens conditioned to various moisture contents. Assembled at 12% m.c., the specimens were conditioned to 16% m.c. and back to 12% m.c. The results of the subsequent tests were, on average, 6% below the test results from the short-term tests for the load-carrying capacity and 10% below the test results for the stiffness values.

Table 7.11: Creep factors for different load durations for single-shear connections with inclined screws and modified surfaces.

Load duration	Time	Creep factor k_{def}		
			Service class 1	
		Connector v1	Connector v2	Mean value
Permanent	50 years	0.8	0.9	0.8
Long term	10 years	0.6	0.7	0.7
Medium term	6 months	0.2	0.4	0.3
			Service class 2	
		Connector v1	Connector v2	Mean value
Permanent	50 years	1.3	1.5	1.4
Long term	10 years	1.1	1.0	1.1
Medium term	6 months	0.5	0.4	0.5

7.3 Connections with laterally loaded fasteners

7.3.1 Duration-of-load tests

7.3.1.1 Test specimens

Tests were carried out with the connections introduced in Chapter 5, Section 5.2 to complement the investigations with the connections with fasteners inserted perpendicular to the shear plane. The investigated surface modifications, however, included only the milled steel plates. The specimens were of the same size as before. For fasteners, 10 mm bolts with large washers (d = 50 mm) were used for each shear plane.

7.3.1.2 Test programme and setup

Three specimens were tested in each service class. The test setup was similar to the short-term test. However, all three specimens were placed in a row, i.e. the lower steel plate of the first specimen was simultaneously the upper steel plate of the second specimen, as was the lower steel plate of the second specimen simultaneously the upper steel plate of the third specimen. The test setup can be seen in Figure 7.19. At the top, the first steel plate was connected with a threaded rod, a washer and a nut to the setup frame. At the bottom, a leverage arm was connected to the last steel plate. The leverage arm was, in turn, supported on one side of the setup frame. The constant load was applied by adding dead weight to the leverage arm. The leverage arm was levelled and checked in intervals. When the leverage arm was out of level, the threaded rod connected to the frame was adjusted, raising or lowering the row of specimens.

The relative displacement between wood and steel was measured with a digital depth gauge. The displacement was measured on the first day of the tests in short intervals, i.e. 10, 15, 30, 45, 60, 90 minutes. In the first week, measurements were taken twice a day. After the first week, measurements were taken twice a week, and after four weeks, measurements were taken once every week. After a period of around five months, measurements were taken twice a month.

(a) (b) (c)

Figure 7.19: Setup of tests with bolts (a+b) and milled steel plate (c).

In Table 7.12, the short-term ultimate load and displacement and the load level for the long-term tests are given. The higher load level of 50% was chosen to compensate for the short duration of the load.

Table 7.12: Overview of long-term tests with connectors with milled surface.

Surface	Fasteners [mm]	$F_{max,mean}$ [kN]	$F_{max,char}$ [kN]	load level (mean)	load level (char.)
Milled circular pattern	1x 10x240	17.1	15.3[1)]	50%	56%

1) 5th-percentile according to EN 14358 with $k_s(n) = 3.15$

7.3.1.3 Results and discussion

Service class 1

The displacements of all three specimens are given in Figure 7.20 in grey, and the averaged displacement in red. The relative displacement increased sharply by about 1 mm in the first few days before the curve turned horizontally. After the first 30 days, there was a sudden increase in deformation, which cannot be explained. After that, the gradient of the curve steadily increased for the remaining time. The displacement seemed to converge at a displacement of 5.5 mm. There was no direct correlation with the environmental conditions, i.e. no sudden increase or decrease of the deformation connected to an increase/decrease of the relative humidity or temperature.

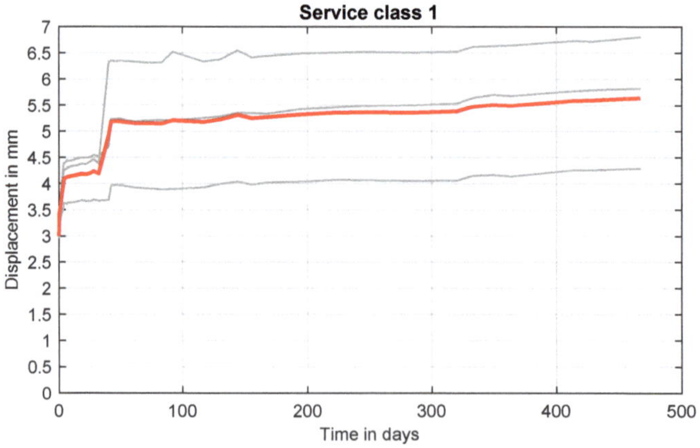

Figure 7.20: Displacement of double-shear steel-to-timber connections in SC1.

The initial displacement after 10 minutes is given in Table 7.13. The connection had an average displacement of 3.19 mm. This result is astonishing, as it was almost twice the deformation measured in the short-term tests (1.69 mm). With this initial displacement, the creep factor k_{def} was calculated.

Table 7.13: Displacement of long-term and short-term tests.

	SC1	SC2	Short-term		
	v in mm	v in mm	v in mm	ρ in kg/m^3	u in %
MEAN	3.19	1.93	1.69	452	10.5
SD	0.69	0.44	0.23	4	0.4
COV	22%	23%	14%	1%	3%

Figure 7.21a shows the creep factor over time. The creep factor for the double-shear steel-to-timber connection was about 0.75 for the test period. The exponential behaviour at the beginning of the tests is visible, as is the steady increase at the end of the tests. Additionally, the simulated creep factors of the ~500 days are given in Figure 7.21a with dashed lines. Again, the models fit the test results quite well.

Figure 7.21b shows the extrapolated curves for 50 years. Here, the models significantly diverge from each other. The creep factor ranges between 0.74–1.18. This divergence again shows the difficulty in simulating the creep factor for 50 years with such a limited database.

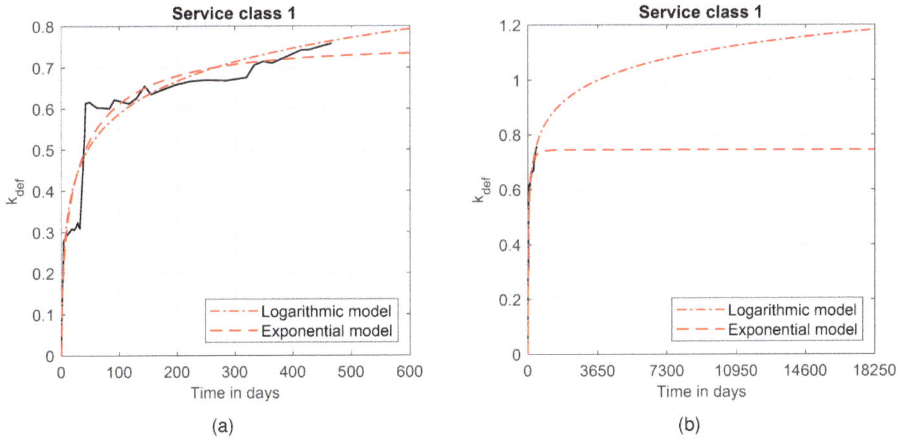

Figure 7.21: Creep factor k_{def} in SC1 for 600 days (a) and 50 years (b) with original displacement data.

The values for the input parameters *a* and *b* are given in Table 7.14. Also given in Table 7.14 are the R^2-values that rate the fit of each model. Both models have a similar R^2-value, which is also higher than for the connections with inclined screws.

Table 7.14: Input parameters for different creep models with original data (SC1).

Model	a	b	R^2
Logarithmic model	0.263	1.718	0.91
Exponential model	0.745	0.172	0.92

The measured displacement in Figure 7.20 showed a sudden increase in deformation, which of course was also visible in the curve of the creep factor in Figure 7.21a. Regardless of the actual origin of this jump, it was assumed that this behaviour was not due to creep. Therefore, the displacement data was adjusted and the creep factor was evaluated one more time. The curves of the creep factor and the two models can be seen in Figure 7.22. With the adjusted data, a creep factor of approximately 0.4 was determined for the test period. With the two models, a creep factor of 0.36–0.55 was determined for the period of 50 years.

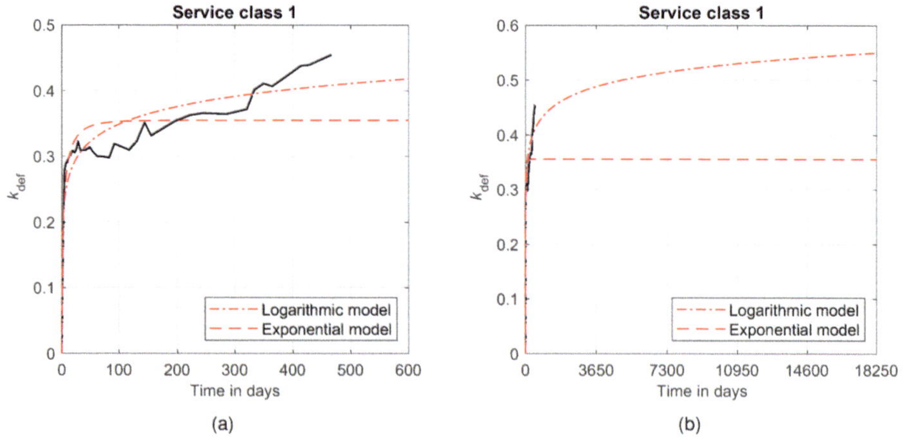

Figure 7.22: Creep factor k_{def} in SC1 for 600 days (a) and 50 years (b) with adjusted displacement data.

The values for the input parameters *a* and *b* are given in Table 7.15. The R^2-values indicate a poorer match of the model with the test data.

Table 7.15: Input parameters for different creep models with adjusted data (SC1).

Model	*a*	*b*	R^2
Logarithmic model	0.089	82.829	0.89
Exponential model	0.355	0.513	0.82

Service class 2

The displacements of the tests in SC2 are given in Figure 7.23. The relative displacement increased constantly for the first 150 days before the curve turned horizontal. During this period, jumps in the deformation were repeatedly recorded, which cannot be explained. After 400 days, there was another sudden increase in deformation of about 3 mm. After that, the gradient of the curve stayed almost horizontal for the remaining time. The displacement seemed to converge at a displacement of 9.5 mm. There was no direct correlation with the environmental conditions, i.e. no sudden increase or decrease of the deformation connected to an increase/decrease of the relative humidity or temperature.

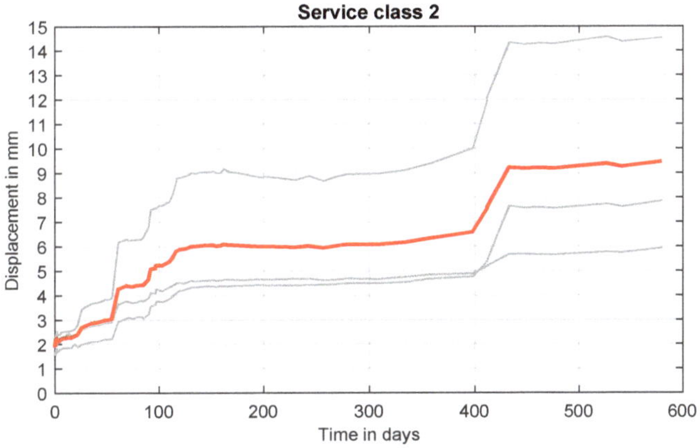

Figure 7.23: Displacement of double-shear steel-to-timber connections in SC2.

The initial displacement after 10 minutes is given in Table 7.13. The connection had an average displacement of 1.93 mm, which fits quite well the short-term tests (1.69 mm). With this initial displacement, the creep factor k_{def} was calculated.

Figure 7.24a plots the creep factor over time. The creep factor for the double-shear steel-to-timber connection is about 3.57 for the test period. The convergence to a horizontal line at the end of the tests is visible. In addition, the extrapolated creep factors of the ~600 days are given in Figure 7.24b with dashed lines. The models fit pretty well with the test results.

Figure 7.24b shows the extrapolated curves for 50 years. Again, the models significantly diverge from each other, with the exponential model still increasing after such a long period. Hence, the creep factor ranges between 8.7 and 19.8.

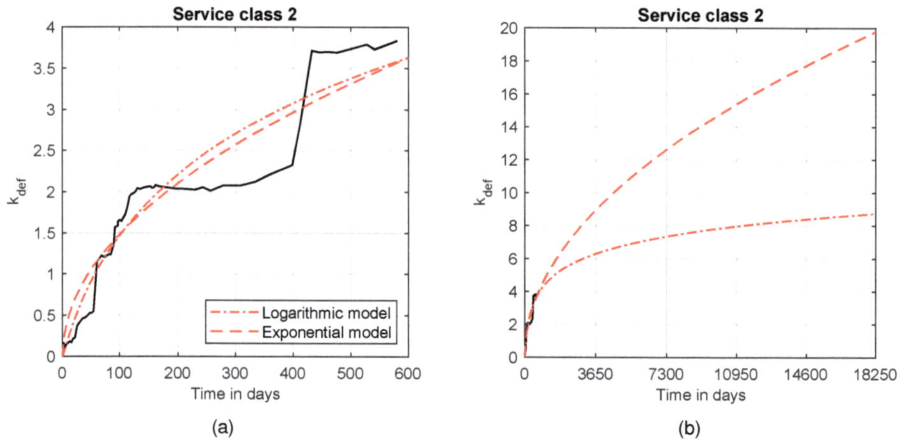

Figure 7.24: Creep factor k_{def} in SC2 for 600 days (a) and 50 years (b) with original displacement data.

The values for the input parameters *a* and *b* are given in Table 7.16, as are the R^2-values. The R^2-values are close to 1.0 and significantly higher than for the tests with inclined screws.

Table 7.16: Input parameters for different creep models with original data (SC2).

Model	*a*	*b*	R^2
Logarithmic model	3.542	0.016	0.96
Exponential model	685	$2.17 \cdot 10^{-4}$	0.95

Again, the data set was adjusted to discount for the jumps in the displacement and hence the creep factor. The plots of the creep factor with the adjusted data are given in Figure 7.25. Both models fit the data quite well, and also for the extrapolation to 50 years, the models both predict similar creep factors. For the test period the creep factor is approximately 1.6 and for the 50 years the creep factor ranges between 2.9 and 3.1.

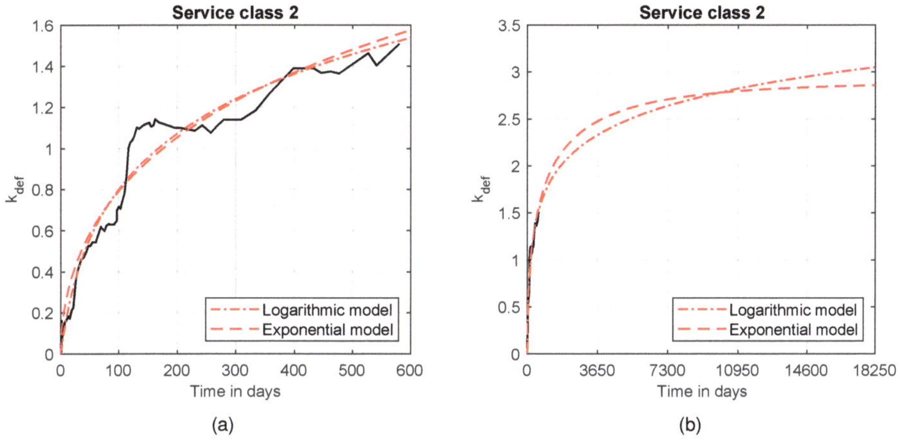

Figure 7.25: Creep factor k_{def} in SC2 for 600 days (a) and 50 years (b) with adjusted displacement data.

The values for the input parameters *a* and *b* are given in Table 7.16. The R^2-values could be further increased with the adjusted displacement data.

Table 7.17: Input parameters for different creep models with adjusted data (SC2).

Model	*a*	*b*	R^2
Logarithmic model	1.030	0.050	0.98
Exponential model	2.896	0.032	0.97

7.3.2 Residual load-carrying capacity

The test specimens in SC 1 were unloaded after 538 days and in SC 2 after 650 days. After unloading, the specimens were left in their respective climatic conditions to recover. The specimens were left to recover for 7 days. The remaining plastic deformation was on average 5.2 mm for SC 1 and 8.3 mm for SC 2. The tests to determine the residual load-carrying capacity were performed analogous to Chapter 5, Section 5.2 and EN 26891 with unloading loop. During the tests, the load and the displacement of the middle and side members were recorded constantly. The displacement was measured with a DIC system and evaluated at the same locations as before in the short-term tests. After the tests, the moisture content and the density were determined individually for each specimen.

Service class 1

The connections with bolts exposed to SC1 reached a mean load of $F_{max} = 12.4 \pm 0.4$ kN. That is only 73% of the mean load of the short-term tests. In Figure 7.26, both the short- and long-term test curves are given. The initial stiffness of the connection is much lower than the short-term tests. This is because of a loss of the preload of the bolts when tightening the nuts. The preload decreases because of creep and shrinking of the timber part. However, after the frictional force is overcome, the connection shows a stiffer behaviour compared to the short-term tests. The stiffness K_s was evaluated in the range of 35–65% of the ultimate load and averaged to 10.3 ± 1.7 kN. This stiffness is almost 3 times higher than the short-term stiffness. At a displacement of about 1.8 mm the slope of the curve changes significantly, and the load barely increases with increasing deformation. Failure of the connections occurs due to splitting of the timber. When the timber starts splitting, the load can be further increased because the washer of the bolts press the timber parts against the steel plate and activate friction. The test results are given in Table 7.18.

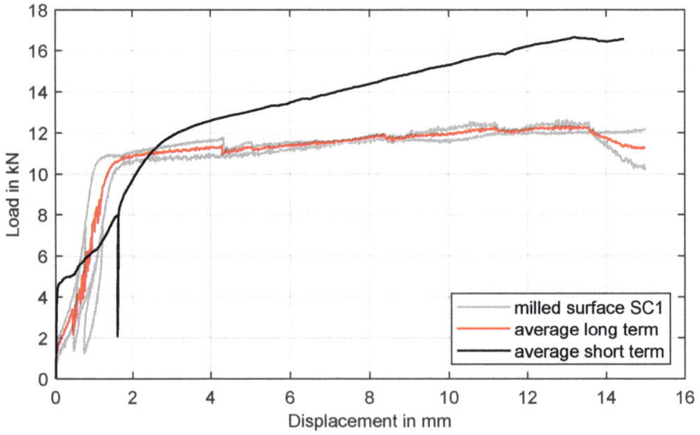

Figure 7.26: Residual load-displacement diagram for tests in SC1.

Table 7.18: Test results of residual load tests (n = 2) in SC1.

	F_{max} [kN]	v_{max} [mm]	K_s [kN/mm]	u [%]
MEAN	12.4	14.0	10.3	10.9
SD	0.3	1.4	1.7	0.2
COV	3%	10%	17%	2%

Figure 7.27 shows the connection after the tests. Splitting of the timber member can be seen. Also clearly visible are wood fibres sticking to the milled surface of the steel plate, showing of the good contact of the milled pattern and the timber, thus increasing the friction in the shear plane. Additionally, it shows that the modified surface was not affected by the changing environmental conditions and the constant load. Figure 7.27b shows the washer of the connection being pressed into the wood. This indentation is the reason for the increasing load, although splitting occurred.

(a) (b)

Figure 7.27: Splitting of wood, and wood fibres on milled surface (a) and indentation of bolt with washer into wood (b).

Service class 2

The tests with the connections conditioned in service class 2 reached a mean load of 18.9 kN, which is 10% higher than the mean load of the short-term tests. The mean stiffness of 40.3±22.4 kN/mm is 12 times higher, compared to the short-term tests. The stiffness K_s was evaluated in the range of 10–40% of the ultimate load. Table 7.19 gives the test results. Figure 7.28 shows the load-displacement diagram. At the beginning, the load-displacement curves are congruent before the curve of the short-term tests flattens significantly due to slip in the connection. The higher stiffness is due to the increased preload of the bolts. The preload of the bolts increased, since swelling of the timber expanded the timber parts. Also, the constant load of approximately 50% of the ultimate short-term load led to significant deformation in the connection and plastic hinges in the bolts.

After reaching a deformation of about 1 mm, the slope decreases and the curve runs parallel to the short-term curve. Failure of the connection occurs due to splitting. The load, however, can still be increased before the tests are terminated at a total displacement of 15 mm. Like the long-term tests with inclined screws, a more humid environment is favourable for the connection, as swelling of the timber in the shear plane is beneficial for the friction in the shear plane. The timber parts are tightly pressed to the steel plate. Because of the modified surface with a milled circular pattern the activated friction in the shear plane is increased.

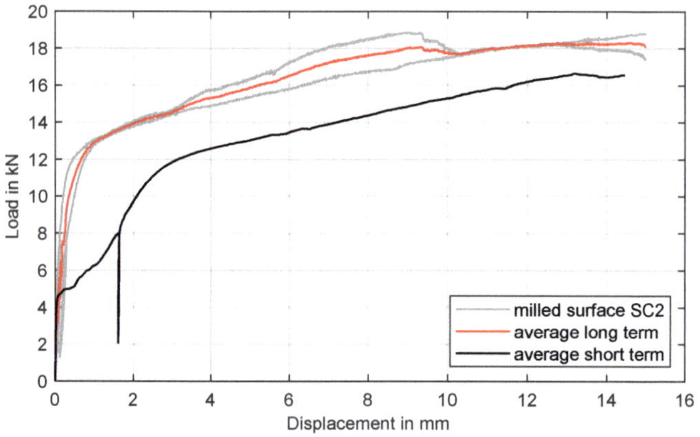

Figure 7.28: Residual load-displacement diagram for tests in SC2.

Table 7.19: Test results of residual load tests (n = 2) in SC2.

	F_{max} [kN]	v_{max} [mm]	K_s [kN/mm]	u [%]
MEAN	18.9	12.0	40.3	15.4
SD	0	4.3	22.4	0.8
COV	0%	35%	56%	5%

Figure 7.29 shows the connection after the tests. Splitting of the timber member can be seen. Also clearly visible are wood fibres sticking to the milled surface of the steel plate, showing of the good contact of the milled pattern and the timber, thus increasing the friction in the shear plane. Additionally, it shows that the modified surface was not affected by the changing environmental conditions and the constant load. Figure 7.29b shows the plastic hinges of the bolts and Figure 7.29c shows the surface of the timber member with the imprint of the milled circular pattern.

(a)

(b)

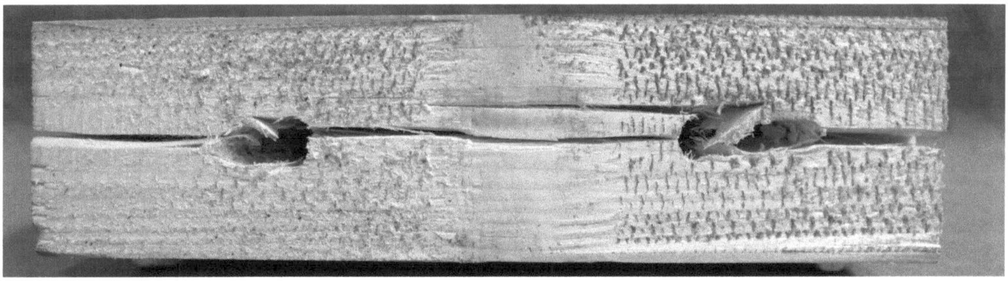

(c)

Figure 7.29: Failure modes: (a) splitting of wood; (b) plastic hinges; (c) imprint of milled circular pattern.

7.3.3 Summary

Duration of load tests were performed with connections with modified surfaces. The tests aimed to investigate the long-term behaviour of connections with bolts inserted perpendicular to the shear plane and increased friction in the shear plane due to a modified surface. A circular pattern was milled into the steel plates for surface modification, and the time of the creep tests covered 470 days in service class 1 and 580 days in service class 2. Creep factors were determined with an initial displacement after 10 minutes. No imminent influence of the temperature on the creep factor is visible and no imminent influence of the relative humidity on the creep factor can be seen. That is, the behaviour of the creep factor curve is not directly linked to the behaviour of the temperature or moisture content curve. The creep factors were extrapolated with different models to a time of 50 years. The results are given in Table 7.20. Here, only the values of the logarithmic model are given. The results of the exponential model were neglected due to the shape of the curve, which seemed unreasonable for both service classes.

Table 7.20: Creep factors for different load durations for connections with bolts inserted perpendicular to the shear plane and modified surface (logarithmic model only).

Load duration	Time	Creep factor k_{def}	
		Service class 1	Service class 2
Permanent	50 years	0.5	3.0
Long term	10 years	0.5	2.3
Medium term	6 months	0.4	1.0

Subsequently, push-out tests were carried out to determine the residual load-carrying capacity of the connections with increased friction in the shear plane. The results showed a decrease in the ultimate load of approx. 25% for service class 1 and an increase of 10% for service class 2. However, the stiffness increased significantly in all tests. For the tests in SC1 the stiffness is 3 times higher, and for the tests in SC2 12 times higher. The failure modes were analogous to the short-term tests, i.e. two plastic hinges per bolt and splitting of the timber.

7.4 Comparison with results from the literature

To categorise the results, they are compared with tests from the literature. These are primarily the experiments by Van de Kuilen [97] with nailed timber-to-timber connections and connections with toothed-plate or split-ring. The test results for connectors v1 and v2 showed a decrease in ultimate load on the one hand but an increase in stiffness on the other. In [97], a ratio is introduced by dividing the results of the residual strength tests (RST) by the results of the standard short-duration (SSD) tests. The ratios are given in Table 7.21. The long-term and short-term strengths ratio is 0.95 and 0.84 for the connectors v1 and v2, respectively. For the stiffness values, an increase was recorded with the long-term tests, and the ratios are 1.13 and 1.07 for the connector v1 and v2, respectively.

With the tested connections in [97], a mixed behaviour of long-term to short-term was observed. While for both the nailed and toothed-plate connections, the long-term strength slightly increased and for the split-ring connection decreased, the stiffness significantly increased for both the toothed-plate and split-ring connections but decreased slightly for the nailed connection. The results of ultimate load and stiffness and the ratio long-term to short-term are given in Table 7.21. The results from literature show no clear tendency for the long-term behaviour of timber connections. However, the results validate the plausibility of the herein-determined ratios.

Table 7.21: Comparison of short-term and long-term results with literature (only service class 1).

	Tested herein			Van de Kuilen [97]		
	inclined screws v1	inclined screws v2	bolts	nails	toothed-plate	split-ring
$F_{\text{max,SSD}}$ [kN]	151	173	17.1	46.1	33.1	28.9
$F_{\text{max.RST}}$ [kN]	143	145	12.4	47.4	35.6	28.5
Ratio RST/SSD	0.95	0.84	0.73	1.03	1.08	0.99
$K_{\text{s.SSD}}$ [kN/mm]	25.0	33.3	3.65	49.5	13.1	27.3
$K_{\text{s.RST}}$ [kN/mm]	28.2	35.7	10.3	45.0	21.0	35.0
Ratio RST/SSD	1.13	1.07	2.82	0.91	1.60	1.28

SSD = standard short duration
RST = residual strength test

The creep factor k_{def} was determined for the same investigated connections. To propose creep factors for the Eurocode 5, the values in [97] were modified to guarantee comparability. This modification is based on the requirement that the expected ultimate deformation has to be equal to the measured ultimate deformation. To be able to compare the results, the creep factors for the connections with inclined screws and with bolts were transformed accordingly. A detailed derivation can be found in [97]. The derived creep factor is as follows:

$$k_{\text{def,EC5}} = \frac{K_{\text{s,exp}}}{K_{\text{s,10min}}} \cdot (1 + k_{\text{def}}) - 1 \tag{7.5}$$

with

$k_{def,EC5}$	proposed creep factor for Eurocode 5
k_{def}	simulated creep factor based on logarithmic/exponential model
$K_{s,exp}$	stiffness of connection according to Eurocode 5
$K_{s,10min}$	stiffness of connection after 10 minutes

The expected stiffness $K_{s,exp}$ was calculated according to the equations given in EC 5. The stiffness after 10 minutes $K_{s,10min}$ was determined with the load of the respective load level and the measured displacement after 10 minutes. For the timber-to-timber connections with fully threaded screws, the average creep factors of connector v1 and v2 from Table 7.11 were used, and for the steel-to-timber connections with bolts, the creep factors from Table 7.20. The values for the other timber-to-timber connections and the concrete-to-timber connections were taken from Van de Kuilen [97] and Van de Kuilen & Dias [98]. The proposed creep factors for connections with inclined screws are about 50% compared to connections with bolts or other dowel-type fasteners. Compared to the other timber-timber connections with different fasteners, the difference is even more significant, with creep factors 3–5 times higher than for connections with inclined screws.

Table 7.22: Comparison of proposed creep factors $k_{def,EC5}$ for different load durations in service class 1.

Load duration	Time	Creep factor $k_{def,EC5}$						
		Timber-timber				Timber-steel	Timber-concrete	
		FTS[1)]	Na[2)]	TP[2)]	SR[2)]	Bo[1)]	DTF[3)]	DTF+[3)]
Permanent	50 years	1.2	4.3	4.4	6.5	2.4	2.1	2.9
Long term	10 years	1.0	3.2	3.6	5.2	2.1	1.9	2.6
Medium term	6 months	0.5	1.1	1.0	1.5	1.6	0.7	0.8

FTS = fully threaded screws
Na = nails
TP = toothed-plates
SR = split-rings
Bo = bolts
DTF = dowel-type fasteners
DTF+ = dowel-type fasteners with interlayer

[1)]tested herein [2)]results from Van de Kuilen [97] [3)]results from Van de Kuilen & Dias [98]

7.5 Conclusion

Tests were carried out in different environmental conditions and varying timber moisture contents. The observed findings suggest that temperature and humidity do not directly influence the deformation of the respective connections, i.e. no direct coupling of the deformation curve to the temperature or humidity curves. However, there is a difference in the deformation when comparing SC1 and SC2. So overall, the different climatic conditions of SC1 and SC2 influence the deformation, but periodical changes within the climatic conditions (i.e. winter and summer) do not directly influence the deformation behaviour. Specifically, connections with inclined screws exhibit a gradual and continuous increase in deformation over an extended period. On the contrary, the bolted connections show an abrupt initial displacement, after which the deformations stabilise relatively swiftly, reaching almost a horizontal equilibrium. This divergence in deformation behaviour implies distinct structural responses, with the connections with inclined screws displaying a prolonged and incremental deformation trend. In contrast, the bolted connections exhibit a more dynamic yet stabilising pattern over time.

When comparing the ratios of short-term ultimate load to long-term ultimate load (or stiffness), it was observed that for the connections with inclined screws, the ultimate load reduced over time, while the stiffness increased slightly. Results from the literature show an inconsistent picture, with certain connections showing a decrease in strength but an increase in stiffness, while in other connections, the strength increases, but the stiffness decreases. Here, it is believed that swelling of the timber enhances the stiffness as possible cavities in the connection are filled. This assumption explains the increased stiffness for the connections with inclined screws and modified surfaces as well as for the connections with toothed-plates and split-rings.

Both models fit the available test data reasonably well. However, the models differ in predicting the creep factor after 50 years. Here, the logarithmic model is more balanced and realistic for timber-to-timber connections, whereas the exponential model is too low or too high. The proposed creep factors for connections with inclined screws are around 50% lower compared to connections with bolts or other dowel-type fasteners. The difference is even more pronounced for timber-to-timber connections using other fasteners, with creep factors 3 to 5 times higher than for connections with inclined screws.

The results of the tests to determine the residual properties show similar behaviour for connections with inclined screws and perpendicular fasteners loaded in shear. For both connection types, the load-carrying capacity decreases, while the stiffness increases.

Also, for both connection types, the results for SC2 are higher than for SC1. This leads to the conclusion that most importantly, a loss of friction is not observed. The decrease in strength is rather due to decreasing mechanical properties (which is considered in the design of connections with the factor k_{mod}).

8 Conclusions and Recommendations

8.1 Conclusions

In the first part of this work, different modification processes were investigated. The aim was to increase the friction coefficient in the shear plane of timber connections. By increasing the friction and, thus, the rope effect, the load-carrying capacity of these connections can be increased.
Friction tests were carried out to determine the coefficient of friction between modified surfaces and softwood. The static and kinetic coefficients of friction were determined for all tests.

In the second part, timber connections with mechanical fasteners were tested. Double-shear steel-to-timber connections with bolts (inserted perpendicular to the shear plane, thus, mainly loaded laterally) and single-shear timber-to-timber connections with screws (inserted inclined at an angle to the shear plane, thus, mainly loaded axially) were investigated. The aim was to identify the influence of friction on the load-carrying capacity and stiffness.
An analytical model was also derived to calculate the load-carrying capacity, while numerical models were derived to determine the deformation of the connections and the stresses in the connectors.

In the last part, the long-term behaviour of the timber connections was investigated. The aim was to investigate the influence of changing environmental conditions. Therefore, the same double-shear steel-to-timber connections with laterally loaded bolts and single-shear timber-to-timber connections with inclined screws like in the short-term tests were used. Creep factors were derived, and the residual load-carrying capacity and stiffness were determined.

Coefficient of friction

Eight different modification processes have been investigated in total, resulting in 16 different variants of surfaces. The surface modifications ranged from simple methods, such as belt grinding and sandblasting, to more complex methods, such as profile milling and coating with sand. Insights into the various surface modification techniques and their feasibility, efficiency, and potential considerations for optimisation in a manufacturing context were provided. Most of the surface modifications were tested with densified veneer wood. These modifications are transferable to other materials, such as steel or aluminium. Some modifications are only feasible with steel or aluminium, and some are only feasible with thin metal sheets.

The tests showed that the effort required to produce the surfaces varied greatly. It became clear that the less complex the production process was, the less consistent the surface quality could be. The investigation also showed that it is difficult to make a statement about the industrial feasibility (ecological and economic), as this was still at the pre-development stage. Appropriate tools would first have to be purchased, which, of course, entails a high investment. In general, the more significant challenge was not to find suitable surfaces but rather the transferability of the surface modification to a suitable application in connections in timber engineering (e.g. punching is not applicable for system connectors).

Nevertheless, it can be concluded from the investigations that surface modification processes such as those classically found in metalworking are to be preferred. The milled surfaces, in particular, produced excellent results for the surface itself and consistent quality during production. Variants with longitudinal and transverse grooves are particularly noteworthy here. Furthermore, it can be concluded from the investigations that a surface treatment in the form of embossing a structure is to be evaluated positively. The surface treatment could be carried out quickly and easily after producing the embossing die. Furthermore, embossing can be used to modify large surfaces easily.

Coatings such as those carried out here are not recommended, as great attention must be paid to the correct choice of adhesive and its processing. It has also been shown that a coating with mineral grains brings significant uncertainty regarding surface quality due to the sometimes widely varying grain sizes of the same coating. If a coating is desired, it is recommended to ensure a very even grain distribution, e.g. corundum or similar.

In total, 26 different surfaces were investigated in friction tests. The tests concluded that all surface treatments led to increasing coefficients of friction compared to the untreated surfaces of DVW or aluminium. The tests also showed that almost all surface modifications led to greater friction coefficients than between wood and wood.

The most valuable conclusion derived from the tests is that large, protruding surface features are unnecessary to achieve a high coefficient of friction in tests. For example, in tests with the embossed surface with the inverse pyramid pattern, convincingly good results of the friction coefficient were determined. There are no protruding features on this surface. Furthermore, the sharper the surface structure, the higher the coefficient of friction that can be determined. Here, the two notched surfaces are a good example.

The tests with the notched surfaces, in particular, showed that the coefficient of friction is limited upwards by the material properties of the wood used. The rolling shear strength, in particular, plays a significant role, as does the compressive strength perpendicular to the grain. The tests were performed with small softwood specimens, and wood failure was decisive. If larger wood specimens are used that better reflect the actual installation state, wood failure should no longer be significant to such an extent, and more realistic friction coefficients should be determined.

The tests also conclude that the surface structure must be distributed as evenly as possible. The coated tests, for example, show that although there is excellent potential due to good interlocking, there are also significant uncertainties due to widely varying grain sizes and their distribution over the surface. This can lead to tilting of the entire surface over individual points.

The first key question could be answered, and surface treatments could be identified that can be produced consistently and significantly increase the coefficient of friction.

Timber connections

Tests were performed with laterally loaded fasteners (inserted perpendicular to the shear plane) and inclined fasteners (mainly loaded axially). In all tests, the surface of the non-timber part (i.e. either steel or DVW) was modified. The results of double-shear steel-to-timber tests with laterally loaded bolts showed only a slight increase in load-carrying capacity (+16–24%) but no increase in stiffness when using a modified steel surface. The results of single-shear DVW-to-timber tests with inclined screws showed a significant increase in the load-carrying capacity of up to 43% (when using a coated surface). The increase in stiffness was not that high, with only +10% on average (however, +27% were reached with an embossed surface).

First of all, it can be concluded from the tests that a surface modification and, thus, an increased friction coefficient in the shear plane increases the load-carrying capacity. This is valid for connections with laterally loaded fasteners inserted perpendicular to the shear plane and connections with fasteners inclined to the shear plane mainly loaded axially. It can also be concluded from the tests that the increase in load-carrying capacity is far more significant than the increase in stiffness.

Here, the advantages of connections with inclined screws are visible. An increase in stiffness between 10–30% was determined in the tests, whereas no increase was found in the tests with the bolts inserted perpendicular to the shear plane. In the tests with the bolts, the increase in load-carrying capacity only occurred with large deformations.

Again, one of the most valuable conclusions that can be derived from the tests is that large, protruding surface structures are not necessary to achieve a high load-carrying capacity and stiffness. The connection tests with connectors with an embossed surface achieved good results. This is explained by the fact that with the "flat" surfaces, the contact pressure is distributed evenly across the shear plane, resulting in more even friction. With the protruding surfaces, more significant displacements are initially required to press the surface completely into the softwood.

With an analytical model, the load-carrying capacity for different surfaces and friction coefficients, as well as for different lengths and numbers of fully threaded screws, was predicted reasonably well. Here, it can be concluded that the static coefficient of friction overestimates the strength, and the kinetic coefficient of friction underestimates the strength of the connection. Therefore, using the static coefficient of friction for the serviceability limit state (SLS) and the kinetic coefficient of friction for the ultimate limit state (ULS) is recommended.

Tests were carried out in different environmental conditions and varying wood moisture contents. The creep factor and the residual strength of the connections were determined. The results show a slightly decreasing load-carrying capacity over time. The portion of the rope effect on the decrease could not be determined. The stiffness, however, could be increased slightly for all tests in both tested service classes. On the other hand, the tests without constant load, where only the moisture content of the timber changed, showed no reduction in strength or stiffness.

The determined creep factors are small for a conditioned environment in service class 1. This goes for both the connection with perpendicular fasteners loaded in shear and the connection with inclined screws. From the tests, it can be concluded that the deformation of connections with inclined screws increases slowly but constantly over time, while the deformation of the connection with perpendicular fasteners loaded in shear increases sharply at the beginning and then stays more or less on the same level over time.

For an unconditioned environment in service class 2, it can be concluded from the tests that the deformation behaviour of the two different connections is similar to the behaviour in service class 1, with a constantly increasing deformation. However, the creep factors are 2–3x as large as in service class 1.

The second key question could also be answered, and the influence of the friction coefficient on the load-carrying capacity and the stiffness could be determined for short-term and long-term behaviour

8.2 Recommendations

Coefficient of friction

Modified surfaces can be used when an increase in load-carrying capacity or fewer fasteners is desired. However, the increase in the friction coefficient cannot be transferred linearly to the increase in load-carrying capacity. When selecting suitable surface modifications, evenly spread features of the surface should be preferred to prominent protruding features.

Further research should be conducted to define a surface modification for large-surface applications. Suitable industrial partners must be identified to investigate and guarantee industrial production.

To determine the coefficient of friction of different surface modifications and timber, a test setup with controllable load application in both directions (parallel and perpendicular to the friction surface) should be chosen. The test setup and testing conditions (such as contact pressure and sliding speed) should be as close as possible to the system behaviour for which the surfaces are tested.

The friction coefficients collected from the literature and the resulting mean values can be used for analytical and numerical modelling and are based on more than 3000 test results. The determined 5%-quantiles of the friction coefficients of the modified surfaces and softwood can be used to design connections if similar surfaces are present.

Further research should consider surface modifications for connections with hardwood. Here, the surface structure presumably has to have sharper, teeth-like features to interlock with the more rigid surface of the hardwood. Subsequently, friction tests must be conducted to determine the friction coefficient between the modified surface and the hardwood. Additionally, surface modifications suited for connections with perpendicular fasteners loaded in shear should be investigated. Here, the surface structure presumably resembles the surface structure for the hardwood connections. This is because the surface has to be pressed into the wood by the contact pressure alone when the screw is screwed in.

Timber connections

An inclined arrangement of the fasteners is preferred when designing connections with dowel-type fasteners and modified surfaces. Due to their immediate loading in the axial direction, the inclined screws were able to press the modified surfaces into the softwood and activate higher friction. The connection should be designed with a decisive failure mode of tensile failure of the screws. For connections with a large number of fasteners on a relatively small area, failure perpendicular to the grain underneath the connector plate must be considered.

Further research should be conducted with single or double-shear connections with laterally loaded fully threaded screws (instead of bolts) inserted perpendicular to the shear plane. On the one hand, it should be checked whether a higher increase in load-carrying capacity can be achieved with screws instead of bolts with washers and nuts. On the other hand, the number and arrangement of the screws should be analysed in the same way as the tests with inclined screws.

An analytical model to calculate the stiffness and deformation of the connection was not investigated. Therefore, an analytical model to determine the stiffness of the connection should be investigated, taking into consideration the increased friction coefficient.

Bibliography

[1] Aurand, S.; Blaß, H.J. (2020): Verbinder aus Kunstharzpressholz mit veränderter Oberfläche zur Erhöhung der Reibung in der Scherfuge. Bautechnik, 97(S1): p. 44–55.

[2] Aurand, S.; Blaß, H.J. (2021): Verbinder aus Kunstharzpressholz – Versuche mit ersten Prototypen für Traglasten bis 500 kN. Bautechnik, 98(S1): p. 40–50.

[3] Aurand, S.; Blaß, H.J. (2023): Verbinder aus Kunstharzpressholz mit erhöhter Reibung in der Scherfuge, Karlsruher Berichte zum Ingenieurholzbau, Volume 38. KIT Scientific Publishing, Karlsruhe.

[4] Aurand, S. (2020): Entwicklung von Hauptträger-Nebenträger-Verbindern aus Kunstharzpressholz: Oberflächenbehandlungen zur Erhöhung der Reibung in der Scherfuge. In 8. Doktorandenkolloquium "Holzbau Forschung + Praxis", Stuttgart.

[5] Aurand, S.; Blaß, H.J. (2021): Development of beam to beam connectors made of densified veneer wood. In World Conference on Timber Engineering WCTE, Santiago, Chile.

[6] Aurand, S.; Blaß, H.J. (2021): Connections with inclined screws and increased shear plane friction. Paper 54-7-5. In Proceedings of the INTER Meeting 54, Online Meeting.

[7] Bejtka, I.; Blaß, H.J. (2002): Joints with Inclined Screws. Paper 35-7-4. In Proceedings of CIB W18 - Meeting 35, Kyoto, Japan.

[8] Gaber, E. (1940): Versuche über die Reibung von Nadelholz. Holz als Roh- und Werkstoff, 3(4): p. 119–122.

[9] Möhler, K.; Maier, G. (1969): Der Reibbeiwert bei Fichtenholz im Hinblick auf die Wirksamkeit reibschlüssiger Holzverbindungen. Holz als Roh- und Werkstoff, 27(8): p. 303–307.

[10] Gečys, T.; Bader, T.K.; Olsson, A.; Kajėnas, S. (2019): Influence of the rope effect on the slip curve of laterally loaded, nailed and screwed timber-to-timber connections. Construction and Building Materials, 228: pp. 116702.

[11] Atack, D.; Tabor, D. (1958): The friction of wood. Proceedings of the Royal Society of London. Series A. Mathematical and Physical Sciences, 246(1247): p. 539–555.

[12] Stošié, D.Z. (1959): Untersuchungen über den statischen Reibungskoeffizienten des Holzes. Holz als Roh- und Werkstoff, 17(3): p. 86–87.

[13] Loo, W.Y.; Quenneville, P.; Chouw, N. (2014): An elastoplastic solution for earthquake resistant rigid timber shear walls. Paper 47-15-6. In Proceedings of the INTER Meeting 47, Bath, United Kingdom.

[14] Zarnani, P.; Valadbeigi, A.; Quenneville, P. (2016): Resilient slip friction (RSF) joint: a novel connection system for seismic damage avoidance design of timber structures. In World Conference on Timber Engineering WCTE, Vienna, Austria.

[15] Johansen, K.W. (1949): Theory of timber connections. International Association for Bridge and Structural Engineering Vol. 9, p. 249–262.

[16] Möller, T. (1951): En ny metod för beräkning av spikförband. Report No. 117. Chalmers University of Technology.

[17] Meyer, A. (1957): Die Tragfähigkeit von Nagelverbindungen bei statischer Belastung. Holz als Roh- und Werkstoff, 15(2): p. 96–109.

[18] Ehlbeck, J. (1979): Nailed joints in wood structures, Wood Research and Wood Construction Laboratory, Volume 166. Virginia Polytechnic Institute and State University, Blacksburg, Virginia.

[19] Bader, T.K.; Schweigler, M.; Serrano, E.; Dorn, M.; Enquist, B.; Hochreiner, G. (2016): Integrative experimental characterization and engineering modeling of single-dowel connections in LVL. Construction and Building Materials, 107: p. 235–246.

[20] Schweigler, M.; Vedovelli, M.; Lemaitre, R.; Bocquet, J.F.; Sandhaas, C.; Bader, T.K. (2021): Beam-on-Foundation modeling as an alternative design method for timber joints with dowel-type fasteners – Part 3: Second order theory effects for considering the rope effect. Paper 54-7-8. In Proceedings of the INTER Meeting 54, Online Meeting.

[21] hardwood_joint (2023): ForestValue: Innovative Verbindungen unter Verwendung von Laubhölzern – Schlussbericht zum FNR-Forschungsvorhaben 22026818.

[22] Domínguez, M.; Fueyo, J.G.; Villarino, A.; Anton, N. (2022): Structural timber connections with dowel-type fasteners and nut-washer fixings: Mechanical characterization and contribution to the rope effect. Materials, 15(1).

[23] Budinski, K.G. (2007): Guide to Friction, Wear and Erosion Testing. ASTM International, West Conshohocken, PA.

[24] Dowson, D. (1979): History of tribology. Longman Group Limited, London.

[25] Gao, C. (1995): Stick-Slip Motion in Boundary Lubrication. Tribology Transactions, 38(2): p. 473–477.

[26] Bowden, F.P. (1958): A review of the friction of solids. Wear, 1(4): p. 333–346.

[27] Blau, P.J. (2001): The significance and use of the friction coefficient. Tribology International, 34(9): p. 585–591.

[28] McLaren, K.G.; Tabor, D. (1961): The frictional properties of lignum vitae. British Journal of Applied Physics, 12(3): p. 118–120.

[29] Murase, Y. (1984): Friction of Wood Sliding on Various Materials. Journal of the Faculty of Agriculture, Kyushu University, 28(4): p. 147–160.

[30] Kuwamura, H. (2011): Coefficient of friction between wood and steel under heavy contact. Journal of Structural and Construction Engineering (Transactions of AIJ), 76(666): p. 1469–1478.

[31] McKenzie, W.M.; Karpovich, H. (1968): The Frictional Behaviour of Wood. Wood Science and Technology, 2(2): p. 139–152.

[32] Lemoine, T.J.; McMillin, C.W.; Manwiller, F.G. (1970): Wood Variables Affecting the Friction Coefficient of Spruce Pine on Steel. Wood Science, 2(3): p. 144–148.

[33] McMillin, C.W.; Lemoine, T.J.; Manwiller, F.G. (1970): Friction Coefficient of Spruce Pine On Steel - A Note on Lubricants. Wood Science, 3(2): p. 100–101.

[34] McMillin, C.W.; Lemoine, T.J.; Manwiller, F.G. (1970): Friction coefficient of oven-dry spruce pine on steel, as related to temperature and wood properties. Wood and Fiber Science, p. 6–11.

[35] Klamecki, B.E. (1976): Friction Mechanisms in Wood Cutting. Wood Science and Technology, 10(3): p. 209–214.

[36] Guan, N.; Thunell, B.; Lyth, K. (1983): On the Friction Between Steel and Some Common Swedish Wood Species. Holz als Roh- und Werkstoff, 41(2): p. 55–60.

[37] McKenzie, W.M. (1991): Friction coefficient as a guide to optimum rake angle in wood machining. Wood Science and Technology, 25: p. 397–401.

[38] Svensson, B.A.; Nyström, S.; Gradin, P.A.; Höglund, H. (2009): Frictional testing of wood — Initial studies with a new device. Tribology International, 42(1): p. 190–196.

[39] Seki, M.; Nakatani, T.; Sugimoto, H.; Miki, T.; Kanayama, K.; Furuta, Y. (2012): Effect of Anisotropy of Wood on Friction Characteristics under High Pressure Conditions. Journal of the Society of Materials Science, Japan, 61(4): p. 335–340.

[40] Seki, M.; Sugimoto, H.; Miki, T.; Kanayama, K.; Furuta, Y. (2012): Wood Friction Characteristics during Exposure to High Pressure: Influence of Moisture Content of Wood. Journal of Mokuzai Gakkaishi, 58(6): p. 302–308.

[41] Seki, M.; Sugimoto, H.; Miki, T.; Kanayama, K.; Furuta, Y. (2013): Wood friction characteristics during exposure to high pressure: influence of wood/metal tool surface finishing conditions. Journal of Wood Science, 59(1): p. 10–16.

[42] Möhler, K.; Herröder, W. (1977): Ermittlung von oberen und unteren Reibungsbeiwertgrenzwerten für den Gleitsicherheitsnachweis bei Traggerüsten (DIN 4421), Bauforschung, Volume T 265. Fraunhofer IRB Verlag, Stuttgart.

[43] Möhler, K.; Herröder, W. (1979): Obere und untere Reibbeiwerte von sägerauhem Fichtenholz. Holz als Roh- und Werkstoff, 37(1): p. 27–32.

[44] Gorst, N.; Williamson, S.J.; Pallett, P.F.; Clark, L.A. (2003): Friction in temporary works: Research report 071. University of Birmingham.

[45] Bejo, L.; Lang, E.M.; Fodor, T. (2000): Friction coefficients of wood-based structural composites. Forest products journal, 50(3): p. 39–43.

[46] Crespo, J.; Regueira, R.; Soilán, A.; Díez, M.R.; Guaita, M. (2011): Desarrollo de metodología para la determinación de los coeficientes de fricción estático y dinámico de diferentes especies de madera. In CIMAD 11 – 1^0 Congresso Ibero-LatinoAmericano da Madeira na Construção, Coimbra, Portugal.

[47] Aira, J.R.; Arriaga, F.; Íñiguez-González, G.; Crespo, J. (2014): Static and kinetic friction coefficients of Scots pine (Pinus sylvestris L.), parallel and perpendicular to grain direction. Materiales de Construcción, 64(315).

[48] Park, C.Y.; Kim, C.K.; Kim, H.K.; Lee, J.J. (2011): Evaluation of Friction Properties According to Normal Force and Direction of Wood Grain in Real Contact Area. Journal of the Korean Wood Science and Technology, 39(5): p. 437–443.

[49] Steiger, R.; Fink, G.; Nerbano, S.; Hack, E.; Beyer, K. (2018): Experimental investigation of friction stresses between adjacent panels made of Oriented Strand Board (OSB) and between OSB panels and glued laminated timber (GLT) frame members. Materials and Structures, 51(1): pp. 123.

[50] Claus, T.; Seim, W.; Liese, J. (2018): Friction under cyclic loading. In World Conference on Timber Engineering WCTE, Seoul, Korea.

[51] Almeida, J.P.; Beyer, K.; Brunner, R.; Wenk, T. (2020): Characterization of mortar–timber and timber–timber cyclic friction in timber floor connections of masonry buildings. Materials and Structures, 53.

[52] Wenk, T.; Brunner, R.; Almeida, J.P.; Beyer, K. (2020): Überprüfung bezüglich Erdbeben von Holzbalkendecken in Bestandsbauten. Bauingenieur, 95(4): p. 8–13.

[53] Meng, Q.; Hirai, T.; Koizumi, A. (2008): Frictional Coefficients between Timber and Some Structural Sheet Materials. Journal of the Japan Wood Research Society, 54(5): p. 281–288.

[54] Hirai, T.; Sawata, K.; Koizumi, A.; Sasaki, Y.; Uematsu, T. (2008): Some Aspects of Frictional Resistance in Timber Construction. In World Conference on Timber Engineering WCTE, Miyazaki, Japan.

[55] Sjödin, J.; Serrano, E.; Enquist, B. (2008): An experimental and numerical study of the effect of friction in single dowel joints. Holz als Roh- und Werkstoff, 66(5): p. 363–372.

[56] Dorn, M. (2012): Investigations on the Serviceability Limit State of Dowel-Type Timber Connections. PhD thesis, Technische Universität Wien, Wien.

[57] Blaß, H.J.; Steige, Y. (2018): Steifigkeit axial beanspruchter Vollgewindeschrauben, Karlsruher Berichte zum Ingenieurholzbau, Volume 34. KIT Scientific Publishing, Karlsruhe.

[58] Dorn, M.; Habrová, K.; Koubek, R.; Serrano, E. (2021): Determination of coefficients of friction for laminated veneer lumber on steel under high pressure loads. Friction, 9(2): p. 367–379.

[59] Koubek, R.; Dedicova, K. (2014): Friction of wood on steel. Master's Thesis, Linnaeus University, Växjö, Sweden.

[60] Rodd, P.D. (1973): The Analysis of Timber Joints Made with Circular Dowel Connectors. PhD thesis, University of Sussex, Sussex.

[61] Schmidt, T. (2018): Kontaktverbindungen für aussteifende Scheiben aus Brettsperrholz, Karlsruher Berichte zum Ingenieurholzbau, Volume 33. KIT Scientific Publishing.

[62] Öniz, S. (2016): Dissipative Stahlblechverbindungen für Brettsperrholz unter wiederholter zyklischer Belastung. Master's thesis (unpublished), Karlsruhe Institute of Technology, Karlsruhe.

[63] Garcia, A. (2012): Untersuchungen zum Tragverhalten von beschichteten HVP-Verbindern. Bachelor's thesis (unpublished), Karlsruhe Institute of Technology, Karlsruhe.

[64] fanabu (2022): Fachwerkträger aus Nadel- und Buchenholz – Schlussbericht zum FNR-Forschungsvorhaben 22010017.

[65] Gressel, P.; Redecker, P. (1991): Reibbeiwerte von Holzwerkstoffen. Forschungsbericht. FH Rosenheim, Rosenheim.

[66] Niemz, P.; Sonderegger, W. (2017): Holzphysik: Physik des Holzes und der Holzwerkstoffe. Carl Hanser Verlag, München.

[67] Koch, H. (2011): Untersuchungen zum Last-Verformungsverhalten historischer Holztragwerke - Der abgestirnte Zapfen, Schriftenreihe Bauwerkserhaltung und Holzbau, Volume 5. Kassel Univ. Press, Kassel.

[68] Xu, M.; Li, L.; Wang, M.; Luo, B. (2014): Effects of Surface Roughness and Wood Grain on the Friction Coefficient of Wooden Materials for Wood–Wood Frictional Pair. Tribology Transactions, 57(5): p. 871–878.

[69] Price, E.; Manwiller, F. (1982): Friction coefficient (on steel) of twenty-two hardwood species, as related to grain orientation. Final Report (Part II) FS-SO-3201-1.45. US Department of Agriculture, Forest Service.

[70] Koch, P. (1985): Utilization of hardwoods growing on southern pine sites. Agriculture Handbook No. 605. US Department of Agriculture, Forest Service.

[71] Schiebel, C. (2018): Experimentelle Ermittlung des Haftreibungskoeffizienten von Kunstharzpressholz. Bachelor's thesis (unpublished), KIT Timber Structures and Building Construction, Karlsruhe Institute of Technology.

[72] Albicker, G. (2023): Experimentelle Ermittlung des Reibbeiwertes von Holz auf verschiedenen Oberflächen. Bachelor's thesis (unpublished), KIT Timber Structures and Building Construction, Karlsruhe Institute of Technology.

[73] Girardon, S.; Bocquet, J.F. (2017): Mechanical behaviour of pre-stressed spruce timber-timber 2.5-mm-step grooved connections under shearing tests. European Journal of Wood and Wood Products, 75(5): p. 719–727.

[74] Jowat Adhesives (2017): Jowat® 2-Component SE Polymer 690.00. Product data sheet. Jowat SE, Detmold, Germany.

[75] Sika (2016): Sikadur®-370 (Provisional). Product data sheet. Sika Services AG, Zurich, Switzerland.

[76] Lohmann Tapes (n.d.): DuploTEC® 10490 SBF - Epoxy Tape. Product data sheet. Lohmann GmbH & Co. KG, Neuwied, Germany.

[77] Lohmann Tapes (n.d.): DuploTEC® 10650 SBF - Epoxy Tape. Product data sheet. Lohmann GmbH & Co. KG, Neuwied, Germany.

[78] Dodge, Y. (2008): The Concise Encyclopedia of Statistics. Springer New York, NY.

[79] Ummenhofer, T.; Boretzki, J.; Albiez, M. (2022): Hybrid connection technologies for hollow sections in steel construction. Steel Construction, 15(S1).

[80] Wursthorn, N. (2022): Experimentelle Untersuchungen zum Verbindungsverhalten von Stahl-Holz-Verbindungen mit verschiedenen Verbindungsmitteln. Bachelor's thesis (unpublished), KIT Timber Structures and Building Construction, Karlsruhe Institute of Technology.

[81] El Hamoui, A. (2019): Experimental studies with connectors composed of laminated densified wood with roughened surfaces and inclined screws. Bachelor's thesis (unpublished), KIT Timber Structures and Building Construction, Karlsruhe Institute of Technology.

[82] Frese, M.; Jordan, M. (2018): Deviations between planned and actual position of wood screws - consequences for minimum spacing. Paper 51-7-1. In Proceedings of the INTER Meeting 51, Tallinn, Estonia.

[83] Blaß, H.J.; Flaig, M. (2019): Blockscheren von Holzbauteilen im Verbindungsbereich axial beanspruchter Vollgewindeschrauben, Karlsruher Berichte zum Ingenieurholzbau, Volume 35. KIT Scientific Publishing, Karlsruhe.

[84] Blaß, H.J. (2017): Selbstbohrende Schrauben und Systemverbinder – Stand der Technik und Herausforderungen. In 23. Internationales Holzbau-Forum IHF, Garmisch-Partenkirchen.

[85] Blaß, H.J.; Bejtka, I.; Uibel, T. (2006): Tragfähigkeit von Verbindungen mit selbstbohrenden Holzschrauben mit Vollgewinde, Karlsruher Berichte zum Ingenieurholzbau, Volume 4. Universitätsverlag Karlsruhe.

[86] Krenn, H. (2018): Die Stahlblech-Holz-Laschenverbindung mit schrägen Schrauben, Monographic Series TU Graz / Timber Engineering & Technology, Volume 7. Verlag der Technischen Universität Graz, Graz.

[87] Frese, M.; Fellmoser, P.; Blaß, H.J. (2010): Modelle für die Berechnung der Ausziehtragfähigkeit von selbstbohrenden Holzschrauben. European Journal of Wood and Wood Products, 68(4): p. 373–384.

[88] Versuchsanstalt für Stahl, Holz und Steine (2019): Prüfbericht Nr. 186147: Ermittlung der mechanischen Eigenschaften von Delignit Panzerholz B15. Unpublished. KIT Timber Structures and Building Construction, Karlsruhe Institute of Technology, Karlsruhe.

[89] Hude, F. (2005): Verbindungssysteme für Queranschlüsse von Neben- an Hauptträger im Ingenieurholzbau. Diploma thesis, Graz University of Technology, Graz.

[90] Proß, C. (2022): Modellierung von Haupt-Nebenträger-Verbinder als Stabwerkssowie 3D-Modell. Master's thesis (unpublished), KIT Timber Structures and Building Construction, Karlsruhe Institute of Technology.

[91] Jockwer, R.; Steiger, R.; Frangi, A. (2014): Design model for inclined screws under varying load to grain angles. Paper 47-7-5. In Proceedings of the INTER Meeting 47, Bath, United Kingdom.

[92] Windeck, L.; Blaß, H.J. (2018): Behaviour of Glulam and LVL Beams Loaded Perpendicular to the Grain. Paper 51-12-3. In Proceedings of the INTER Meeting 51, Tallinn, Estonia.

[93] Borst, J. (2019): FE-Modellierung des Tragverhaltens von Systemverbindern für Hauptträger-Nebenträger-Verbindungen mit axial beanspruchten Vollgewindeschrauben. Master's thesis (unpublished), KIT Timber Structures and Building Construction, Karlsruhe Institute of Technology.

[94] Heisig, L. (2019): FE-Modellierung von Verbindern aus Kunstharzpressholz mit axial beanspruchten Vollgewindeschrauben. Master's thesis (unpublished), KIT Timber Structures and Building Construction, Karlsruhe Institute of Technology.

[95] Avez, C.; Descamps, T.; Serrano, E.; Léoskool, L. (2016): Finite element modelling of inclined screwed timber to timber connections with a large gap between the elements. European Journal of Wood and Wood Products, 74(3): p. 467–471.

[96] Azinović, B.; Frese, M. (2020): FE modelling of timber connections with inclined and cross-wise arranged screws - new findings on testing the shear stiffness. Paper 53-7-2. In Proceedings of the INTER Meeting 53, Online Meeting.

[97] Van de Kuilen, J.W.G. (1999): Duration of Load Effects in Timber Joints. PhD thesis, Delft University of Technology, Delft, The Netherlands.

[98] Van de Kuilen, J.W.G.; Dias, A. (2015): Creep factors for timber-timber and timber-concrete joints. In Studies and Researches, Volume 34, Graduate School in Concrete Structures – Fratelli Pesenti, Politecnico di Milano, Milan, Italy.

[99] Pfefferle, R. (1971): Zur Theorie des Betonkriechens. PhD thesis, Universität Karlsruhe, Fakultät für Bauingenieur- und Vermessungswesen.

[100] Reinhardt, H. (1973): Zur Beschreibung des rheologischen Verhaltens von Holz. Holz als Roh- und Werkstoff, 31(9): p. 352–355.

[101] Gräfe, M. (2020): Vorgespannte Konstruktionen aus Brettsperrholz – Entwicklung, experimentelle und theoretische Untersuchungen, Entwurf und Bemessung. PhD thesis, Technische Universität München, München.

[102] Blaß, H.J.; Sandhaas, C. (2017): Timber Engineering - Principles for design. KIT Scientific Publishing, Karlsruhe.

[103] Steck, G. (1982): Die Zuverlässigkeit des Vollholzbalkens unter reiner Biegung. PhD thesis, Universität Karlsruhe, Fakultät für Bauingenieur- und Vermessungswesen.

[104] Blaß, H.J. (2018): Moment-Normalkraft-Querkraft Interaktion in stiftförmigen Verbindungsmitteln von Stahlblech-Holz-Verbindungen. In Karlsruher Tage 2018 Holzbau, Karlsruhe.

Standards and approvals

[105] ASTM Standard G115-10 (2018) Guide for Measuring and Reporting Friction Coefficients. ASTM International, West Conshohocken, PA.

[106] DIN 1052 (2004) Design of timber structures – General rules and rules for buildings. DIN Deutsches Institut für Normung e.V., Berlin.

[107] DIN 8580 (2022) Manufacturing processes – Terms and definitions, division. DIN Deutsches Institut für Normung e.V., Berlin.

[108] DIN 8583-5 (2003) Manufacturing processes forming under compressive conditions – Part 5: Indentation-forming – Classification, subdivision, terms and. DIN Deutsches Institut für Normung e.V., Berlin.

[109] DIN 8589-12 (2003) Manufacturing processes chip removal – Part 12: Belt grinding (sanding) – Classification, subdivision, terms and definitions. DIN Deutsches Institut für Normung e.V., Berlin.

[110] EN 12812 (2008) Falsework – Performance requirements and general design; German version. DIN Deutsches Institut für Normung e.V., Berlin.

[111] EN 14358 (2016) Timber structures – Calculation and verification of characteristic values; German version. DIN Deutsches Institut für Normung e.V., Berlin.

[112] EN 14545 (2009) Timber structures – Connectors – Requirements; German version. DIN Deutsches Institut für Normung e.V., Berlin.

[113] EN 14592 (2012) Timber structures – Dowel-type fasteners — Requirements; German version. DIN Deutsches Institut für Normung e.V., Berlin.

[114] EN 1995-1-1 (2012) Eurocode 5: Design of timber structures – Part 1-1: General – Common rules and rules for buildings; German version. DIN Deutsches Institut für Normung e.V., Berlin.

[115] EN 1995-2 (2010) Eurocode 5: Design of timber structures – Part 2: Bridges; German version. DIN Deutsches Institut für Normung e.V., Berlin.

[116] EN 26891 (1991) Timber structures; Joints made with mechanical fasteners; General principles for the determination of strength and deformation characteristic (ISO 6891:1983). DIN Deutsches Institut für Normung e.V., Berlin.

[117] ETA-11/0190 (2018) Würth self-tapping screws. Deutsches Institut für Bautechnik (DIBt), Berlin.

[118] ETA-19/0553 (2021) HECO-TOPIX-plus (or HTP or HT-plus), HECO-TOPIX-plus-T (or HTP-T or HT-plus-T) and HECO-TOPIX-plus-CC screws (or HTP-CC or HT-plus-CC) Screws for use in timber constructions. ETA-Danmark A/S.

Software

[119] The MathWorks (2023): MATLAB R2023b. The MathWorks, Inc., Natick, MA, USA.

[120] SAS Institute (2016): SAS 9.4. SAS Institute Inc., Cary, NC, USA.

[121] Dassault Systèmes (2019): Abaqus/CAE 2020. Dassault Systèmes Simulia Corp., Johnston, RI, USA.

Acronyms and symbols

Acronyms

ANOVA	analysis of variance
Bo	bolts
CDF	cumulative distribution function
CLT	cross laminated timber
COF	coefficient of friction
COV	coefficient of variation
DIC	digital image correlation
DTF	dowel-type fastener
DVW	densified veneer wood
FTS	fully threaded screw
GLT	glued laminated timber
KS	Kolmogorov-Smirnov
MAX	maximum value
MEAN	mean value
MIN	minimum value
Na	nails
RST	residual strength test
SC	service class
SD	standard deviation
SSD	standard short duration
SR	split-ring
TP	toothed-plate

Latin symbols and variables

$A_{c,90,H}$	area under the connector plate at the header
$A_{net,DVW}$	net cross-sectional area of the DVW connector
a	parameter for creep model determined from test results
b	parameter for creep model determined from test results
d	diameter of fastener
d_{head}	head diameter of fastener
F_{ax}	withdrawal capacity of fastener
$F_{ax,Rk}$	characteristic load-carrying capacity of fastener
F_{f}	frictional force parallel to the shear plane
$F_{max,SSD}$	ultimate load in standard short duration (short-term) tests
$F_{max,RST}$	ultimate load in residual strength tests
F_{n}	normal force perpendicular to the shear plane
F_{tens}	tensile capacity of fastener
$F_{V,test}$	ultimate (test) load from experimental test
$F_{V,exp}$	ultimate (expected) load from analytical model
$f_{c,0}$	compression strength parallel to the grain
$f_{c,90}$	compression strength perpendicular to the grain
$f_{h,i}$	embedment strength
$k_{c,90}$	coefficient for compression perpendicular to the grain
$k_{def,con}$	creep factor of connection
$k_{def,EC5}$	proposed creep factor for EC 5
K_{i}	stiffness in the range of 0–40% of ultimate load
K_{s}	stiffness in the range of 10–40% of ultimate load
$K_{s,10min}$	stiffness of connection after 10 minutes long-term test
$K_{s,exp}$	expected stiffness of connection according to EC 5
K_{SLS}	mean slip modulus (stiffness) of a connection
$K_{SLS,fin}$	final mean slip modulus (stiffness)
ℓ	length of fastener

M_y	bending moment of fastener
m.c.	moisture content
n	number of fasteners
n_{ef}	effective number of fasteners
t	time in days
R	load-carrying capacity of connection
u	moisture content
u	measured deformation of a connection
u_{inst}	initial deformation of a connection
v	testing speed in friction tests
v_{max}	displacement of connection at ultimate load
v_{10min}	displacement of connection after 10 minutes

Greek symbols and variables

β	ratio of $f_{h,1}$ to $f_{h,2}$
ε	insertion angle of screw in regards to the shear plane
μ_s	static coefficient of friction
μ_k	kinetic coefficient of friction
ρ	density
σ_N	contact pressure in the shear plane

A Appendix

A.1 Friction tests

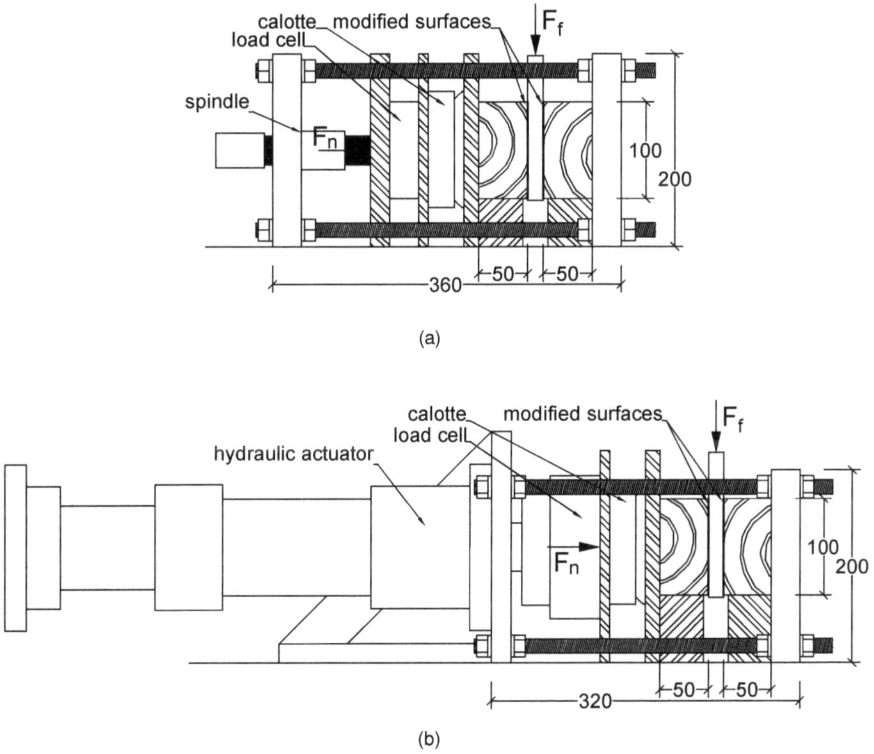

Figure A.1: Test setups for friction tests with (a) spindle and (b) hydraulic actuator.

Table A.1: Friction coefficients for all tested surfaces.

Test	Face grain $\parallel$					Face grain $\perp$					End grain				
	σ_N	v	ρ	μ_{stat}	μ_{kin}	σ_N	v	ρ	μ_{stat}	μ_{kin}	σ_N	v	ρ	μ_{stat}	μ_{kin}
Untreated – densified veneer wood (DVW) and softwood															
1	0.50	1	428	0.20	0.19	1.14	1	-	0.18	0.19	1.16	1	-	0.18	0.18
2	0.50	1	494	0.25	0.22	1.11	1	-	0.18	0.19	1.03	1	-	0.19	0.18
3	0.50	1	504	0.21	0.20	1.09	1	-	0.18	0.19	1.15	1	-	0.18	0.18
4	0.50	1	434	0.21	0.19	1.40	5	-	0.22	0.20	1.10	5	-	0.19	0.19
5	0.50	1	400	0.21	0.19	1.54	5	-	0.20	0.20	1.15	5	-	0.22	0.19
6	0.50	5	456	0.23	0.20	1.30	5	-	0.18	0.20	1.18	5	-	0.21	0.19
7	0.50	5	445	0.26	0.22	1.18	10	-	0.17	0.16	1.13	10	-	0.16	0.14
8	0.50	5	427	0.26	0.20	1.06	10	-	0.16	0.16	1.19	10	-	0.17	0.17
9	0.50	5	462	0.23	0.21	1.14	10	-	0.17	0.17	1.10	10	-	0.21	0.19
10	0.50	5	477	0.18	0.17	2.24	1	-	0.16	0.19	2.57	1	-	0.19	0.18
11	0.50	10	441	0.21	0.18	2.47	1	-	0.22	0.20	2.69	1	-	0.24	0.20
12	0.50	10	454	0.23	0.20	2.38	1	-	0.27	0.26	2.66	1	-	0.20	0.18
13	0.50	10	437	0.19	0.18	2.46	1	-	0.25	0.26	2.49	5	-	0.21	0.18
14	0.50	10	417	0.20	0.18	2.57	5	-	0.23	0.18	2.58	5	-	0.17	0.16
15	0.50	10	471	0.23	0.20	2.51	5	-	0.19	0.18	2.72	5	-	0.18	0.17
16	1.00	1	434	0.25	0.21	2.37	5	-	0.24	0.22	2.63	10	-	0.18	0.17
17	1.00	1	453	0.23	0.19	2.49	10	-	0.25	0.23	2.66	10	-	0.21	0.19
18	1.00	1	419	0.22	0.19	2.66	10	-	0.20	0.20	2.66	10	-	0.21	0.20
19	1.00	5	465	0.22	0.19	2.65	10	-	0.20	0.20	5.94	1	-	0.17	0.16
20	1.00	5	449	0.21	0.19	5.71	1	-	0.19	0.16	6.12	1	-	0.16	0.16
21	1.00	5	470	0.22	0.21	5.21	1	-	0.20	0.17	6.33	1	-	0.16	0.17
22	1.00	10	432	0.27	0.22	5.98	1	-	0.22	0.19	6.08	5	-	0.17	0.17
23	1.00	10	430	0.27	0.23	6.05	1	-	0.21	0.17	6.38	5	-	0.16	0.17
24	1.00	10	469	0.25	0.21	5.97	5	-	0.20	0.17	6.54	5	-	0.16	0.18
25	2.50	1	424	0.21	0.19	5.72	5	-	0.20	0.18	6.31	10	-	0.19	0.17
26	2.50	1	456	0.22	0.20	5.95	5	-	0.20	0.18	6.36	10	-	0.18	0.17
27	2.50	1	412	0.20	0.18	6.27	5	-	0.19	0.17	6.16	10	-	0.19	0.19
28	2.49	5	419	0.23	0.19	5.96	10	-	0.18	0.16	0.25	1	378	0.19	0.20
29	2.51	5	422	0.28	0.21	5.66	10	-	0.18	0.16	0.25	1	392	0.20	0.19

continued on next page

Table A.1 – continued from previous page

Test	Face grain ∥					Face grain ⊥					End grain				
	σ_N	v	ρ	μ_{stat}	μ_{kin}	σ_N	v	ρ	μ_{stat}	μ_{kin}	σ_N	v	ρ	μ_{stat}	μ_{kin}
30	2.50	5	417	0.26	0.21	6.07	10	-	0.19	0.17	0.25	1	396	0.23	0.20
31	2.50	10	465	0.22	0.18	6.27	10	-	0.18	0.17	0.25	1	-	0.17	0.17
32	2.50	10	453	0.25	0.21	0.50	1	470	0.31	0.22	0.25	1	-	0.19	0.20
33	2.50	10	459	0.25	0.22	0.50	1	459	0.34	0.25	0.25	5	378	0.20	0.17
34	6.00	1	456	0.19	0.17	0.50	1	457	0.33	0.24	0.25	5	392	0.25	0.20
35	6.00	1	510	0.21	0.19	0.50	1	462	0.28	0.20	0.25	5	396	0.21	0.20
36	6.00	1	416	0.18	0.17	0.50	1	452	0.42	0.34	0.25	5	-	0.23	0.21
37	6.00	5	468	0.20	0.18	0.50	5	476	0.27	0.20	0.25	5	-	0.19	0.18
38	6.00	5	424	0.18	0.17	0.50	5	461	0.29	0.21	0.25	10	492	0.22	0.22
39	6.00	5	432	0.19	0.16	0.50	5	454	0.31	0.21	0.25	10	492	0.21	0.21
40	5.98	10	476	0.20	0.19	0.50	5	473	0.31	0.22	0.25	10	476	0.17	0.15
41	6.00	10	439	0.19	0.17	0.50	5	450	0.38	0.26	0.25	10	-	0.14	0.15
42	6.00	10	424	0.19	0.18	0.50	10	455	0.25	0.19	0.25	10	-	0.21	0.19
43						0.50	10	468	0.29	0.21	0.50	1	396	0.17	0.17
44						0.50	10	435	0.32	0.25	0.50	1	398	0.18	0.17
45						0.50	10	447	0.33	0.23	0.50	1	424	0.23	0.23
46						0.50	10	410	0.33	0.21	0.50	1	388	0.20	0.20
47						1.00	1	478	0.30	0.22	0.50	1	416	0.26	0.20
48						1.00	1	421	0.26	0.19	0.50	5	420	0.19	0.19
49						0.99	1	427	0.29	0.20	0.50	5	450	0.21	0.22
50						1.01	5	423	0.20	0.17	0.50	5	451	0.21	0.19
51						0.99	5	443	0.28	0.20	0.50	5	451	0.23	0.20
52						1.00	5	472	0.28	0.20	0.50	5	407	0.23	0.19
53						1.00	10	457	0.22	0.21	0.50	10	417	0.23	0.24
54						1.00	10	437	0.33	0.25	0.50	10	413	0.18	0.16
55						1.00	10	444	0.27	0.20	0.50	10	417	0.19	0.17
56						2.50	1	454	0.18	0.17	0.50	10	460	0.27	0.25
57						2.50	1	448	0.24	0.19	0.50	10	419	0.23	0.23
58						2.50	1	439	0.22	0.18	1.00	1	417	0.24	0.30
59						2.50	5	437	0.18	0.17	1.00	1	432	0.20	0.20
60						2.50	5	460	0.21	0.17	1.00	1	450	0.18	0.17

continued on next page

Table A.1 – continued from previous page

Test	Face grain ∥					Face grain ⊥					End grain				
	σ_N	v	ρ	μ_{stat}	μ_{kin}	σ_N	v	ρ	μ_{stat}	μ_{kin}	σ_N	v	ρ	μ_{stat}	μ_{kin}
61						2.49	5	444	0.24	0.17	1.00	5	482	0.29	0.31
62						2.51	10	465	0.25	0.20	1.00	5	421	0.23	0.22
63						2.50	10	436	0.29	0.23	1.00	5	413	0.24	0.23
64						2.49	10	465	0.24	0.18	1.00	10	438	0.26	0.24
65						6.00	1	462	0.20	0.17	1.00	10	447	0.20	0.17
66						6.00	1	442	0.17	0.15	1.00	10	427	0.21	0.21
67						6.00	1	452	0.17	0.15	6.00	1	412	0.15	0.16
68						6.04	5	478	0.18	0.16	6.00	1	432	0.18	0.21
69						6.00	5	482	0.21	0.17	6.00	1	384	0.17	0.16
70						5.99	5	446	0.19	0.16	6.00	5	464	0.17	0.16
71						6.00	10	434	0.19	0.16	5.99	5	388	0.16	0.16
72						5.99	10	442	0.18	0.15	5.99	5	440	0.17	0.16
73						6.00	10	430	0.18	0.16	6.00	10	398	0.18	0.18
74											6.00	10	416	0.18	0.19
75											6.00	10	376	0.22	0.19
76											2.50	1	440	0.22	0.22
77											2.50	1	464	0.23	0.21
78											2.50	1	384	0.21	0.19
79											2.50	5	485	0.17	0.17
80											2.50	5	481	0.18	0.19
81											2.50	5	432	0.18	0.19
82											2.50	10	456	0.19	0.16
83											2.50	10	423	0.18	0.17
84											2.50	10	444	0.21	0.24
MEAN			446	0.22	0.19			451	0.23	0.19			427	0.20	0.19
COV			6%	12%	8%			4%	25%	17%			7%	15%	15%
Untreated – aluminium and softwood															
1	2.53	10	436	0.32	0.29	2.22	5	-	0.30	0.28	1.26	5	-	0.35	0.44
2	2.54	10	436	0.21	0.17	2.22	5	-	0.30	0.30	1.23	5	-	0.40	0.43
3	2.53	10	439	0.18	0.18	2.60	5	-	0.30	0.27	1.21	5	-	0.32	0.44
4	2.53	10	-	0.18	0.20	2.12	5	-	0.38	0.47	1.19	5	-	0.35	0.43

continued on next page

Table A.1 – continued from previous page

Test	Face grain ∥					Face grain ⊥					End grain				
	σ_N	v	ρ	μ_{stat}	μ_{kin}	σ_N	v	ρ	μ_{stat}	μ_{kin}	σ_N	v	ρ	μ_{stat}	μ_{kin}
5	2.54	10	-	0.21	0.18	2.44	5	-	0.42	0.57	1.28	5	-	0.31	0.43
6	2.54	10	-	0.31	0.34	2.37	5	-	0.37	0.51	1.28	5	-	0.44	0.47
7	2.53	10	502	0.25	0.24	1.04	5	-	0.28	0.35	1.20	5	-	0.37	0.46
8	2.53	10	502	0.33	0.33	0.93	5	-	0.37	0.43	1.22	5	-	0.46	0.47
9	2.54	10	522	0.26	0.25	0.82	5	-	0.46	0.54	1.21	5	-	0.44	0.46
10	2.53	10	540	0.28	0.28	1.12	5	-	0.28	0.36	1.13	5	-	0.38	0.45
11	2.53	10	-	0.27	0.24	1.20	5	-	0.31	0.42	1.20	5	-	0.43	0.47
12	2.54	10	-	0.26	0.22	1.14	5	-	0.34	0.43	1.20	5	-	0.37	0.47
MEAN			482	0.25	0.24				0.34	0.41				0.39	0.45
COV			9%	20%	23%				17%	24%				13%	4%
Notched – file cut and softwood															
1	1.1	5	445	1.76	0.62	1.1	5	417	1.04	0.58	1.1	5	450	0.88	0.64
2	1.1	5	472	2.08	0.63	1.1	5	469	0.79	0.71	1.1	5	456	0.94	0.80
3	1.1	5	456	1.88	0.69	1.1	5	432	1.11	0.58	1.1	5	423	1.55	0.96
4	1.1	5	457	2.04	0.65	1.1	5	483	1.14	0.61	1.1	5	421	0.95	0.94
5	1.1	5	427	1.44	0.48	1.1	5	434	1.18	0.62	1.1	5	417	1.15	1.09
6	1.1	5	437	2.08	0.56	1.1	5	430	1.12	0.63	1.1	5	460	0.94	0.89
7	1.1	5	471	1.40	0.61	1.1	5	422	1.10	0.51	1.1	5	419	1.34	0.87
8	1.1	5	417	1.94	0.60	1.1	5	465	1.01	0.62	1.1	5	485	1.12	0.67
MEAN			448	1.83	0.60			444	1.06	0.61			441	1.11	0.86
COV			4%	15%	10%			6%	11%	9%			6%	21%	18%
Notched – rasp cut and softwood															
1	1.0	5	438	1.94	0.71	1.0	5	437	0.86	0.69	1.0	5	447	1.47	1.41
2	1.0	5	476	2.43	0.66	1.0	5	465	1.23	0.72	1.0	5	413	1.48	1.09
3	1.0	5	446	2.40	0.66	1.0	5	442	1.27	0.79	1.0	5	427	1.75	1.05
4	1.0	5	455	2.28	0.64	1.0	5	452	1.24	0.79	1.0	5	420	1.37	1.31
5	0.9	5	461	2.41	0.65	1.0	5	421	0.94	0.63	0.9	5	452	1.33	1.48
6	1.0	5	450	2.50	0.59	1.0	5	410	0.95	0.67	1.0	5	407	1.44	1.55
7	1.0	5	440	1.84	0.50	1.0	5	457	1.39	0.74	1.0	5	417	1.45	1.37
8	1.0	5	465	2.74	0.69	1.0	5	447	1.26	0.76	1.0	5	438	1.46	1.70

continued on next page

Table A.1 – continued from previous page

Test	Face grain ‖					Face grain ⊥					End grain				
	σ_N	v	ρ	μ_{stat}	μ_{kin}	σ_N	v	ρ	μ_{stat}	μ_{kin}	σ_N	v	ρ	μ_{stat}	μ_{kin}
9	0.4	5	-	2.71	0.73	0.4	5	-	2.18	0.76					
10	0.4	5	-	3.81	1.06	0.4	5	-	1.45	0.80					
11	0.4	5	-	3.58	0.73	0.4	5	-	1.89	0.75					
MEAN			454	2.60	0.69			441	1.33	0.74			427	1.47	1.37
COV			3%	23%	20%			4%	30%	7%			4%	8%	16%
Embossed – inverse pyramid pattern and softwood															
1	2.07	5	439	0.69	0.52	2.80	5	443	0.67	0.70	3.51	5	405	0.77	0.57
2	2.24	5	442	0.70	0.56	2.05	5	446	0.74	0.50	3.78	5	407	0.78	0.69
3	2.26	5	440	0.68	0.47	2.67	5	445	0.62	0.48	5.27	5	409	0.64	0.53
4	2.10	5	437	0.65	0.52	2.00	5	446	0.54	0.31	5.05	5	431	0.72	0.53
5	2.19	5	432	0.87	0.59	2.10	5	449	0.76	0.44	4.97	5	435	0.74	0.52
6	2.46	5	436	0.76	0.51	2.15	5	454	0.86	0.65	4.33	5	474	0.68	0.62
7	2.16	5	429	0.65	0.46	2.46	5	458	0.79	0.52	5.26	5	522	0.67	0.47
8	2.54	5	428	0.57	0.44	1.95	5	467	0.87	0.40	5.59	5	530	0.62	0.57
9	2.25	5	422	0.66	0.53	2.28	5	475	0.86	0.51	5.09	5	533	0.59	0.54
10	2.42	5	426	0.54	0.41	1.66	5	477	0.91	0.66	5.42	5	546	0.68	0.55
11						2.34	5	385	0.75	0.47	5.54	5	548	0.69	0.53
12						2.16	5	399	0.77	0.60	1.60	5	383	0.76	-
13						2.11	5	381	0.82	0.49	1.69	5	384	0.72	0.38
14						1.81	5	382	0.90	0.51	1.77	5	389	0.78	0.54
15						1.91	5	483	1.05	0.73	1.60	5	401	0.77	0.46
16						1.97	5	506	0.84	0.62	1.70	5	457	0.77	0.51
17						2.54	5	494	0.96	0.66	1.82	5	467	0.64	0.46
18						2.22	5	493	0.78	0.69	1.89	5	469	0.70	0.46
19						2.09	5	487	0.81	0.64	2.00	5	489	0.75	0.46
20						2.27	5	393	0.75	0.43	2.08	5	574	0.68	0.49
21						4.94	5	407	0.75	0.40	2.41	5	597	0.80	0.53
22						4.87	5	408	0.77	0.55					
23						3.64	5	410	0.78	0.49					
24						5.15	5	413	0.75	0.56					
25						4.59	5	416	0.70	0.48					

continued on next page

Table A.1 – continued from previous page

Test	Face grain ‖					Face grain ⊥					End grain				
	σ_N	v	ρ	μ_{stat}	μ_{kin}	σ_N	v	ρ	μ_{stat}	μ_{kin}	σ_N	v	ρ	μ_{stat}	μ_{kin}
26						5.79	5	552	0.67	0.36					
27						4.70	5	547	0.68	0.33					
28						6.58	5	563	0.92	0.44					
29						4.36	5	559	0.82	0.42					
30						4.19	5	556	0.89	0.47					
MEAN			433	0.68	0.50			460	0.79	0.52			469	0.71	0.52
COV			2%	14%	11%			12%	13%	22%			14%	8%	13%
Embossed – chequered plate and softwood															
1	1.92	5	-	0.89	0.85	1.56	5	-	0.99	1.00	1.25	5	376	0.55	0.59
2	1.96	5	-	1.05	0.90	2.01	5	-	0.70	0.70	1.25	5	416	0.61	0.66
3	1.91	5	-	0.86	0.84	1.89	5	-	0.81	0.83	1.25	5	398	0.56	0.61
4	0.92	5	-	0.71	0.63	1.25	5	-	0.70	0.73	1.25	5	388	0.64	0.67
5	0.92	5	-	0.80	0.75	1.16	5	-	0.62	0.58	1.25	5	412	0.58	0.63
6	0.95	5	-	0.76	0.73	1.21	5	-	0.62	0.61	1.25	5	440	0.58	0.59
7	2.50	5	438	0.62	0.67	2.50	5	435	0.40	0.46	1.25	5	404	0.66	0.71
8	2.50	5	406	0.51	0.56	2.50	5	457	0.51	0.51	2.50	5	452	0.59	0.65
9	2.50	5	488	0.50	0.56	2.50	5	480	0.54	0.57	2.50	5	451	0.55	0.62
10	2.50	5	387	0.47	0.50	2.50	5	539	0.50	0.57	2.50	5	454	0.57	0.62
11	2.50	5	495	0.54	0.57	2.50	5	397	0.44	0.42	2.50	5	459	0.57	0.64
12	2.50	5	480	0.43	0.49	2.50	5	459	0.44	0.45	2.50	5	467	0.60	0.68
13	2.50	5	398	0.41	0.40	2.50	5	370	0.41	0.48	2.50	5	464	0.61	0.69
MEAN			442	0.66	0.65			448	0.59	0.61			429	0.59	0.64
COV			10%	31%	24%			12%	30%	28%			7%	6%	6%
Punched – perforation 1 and softwood															
1									0.69	0.40					
2									0.81	0.48					
3									0.95	0.59					
MEAN									0.82	0.49					
COV									16%	19%					
Punched – perforation 2 and softwood															
1	2.5	5	438	0.61	0.44	2.5	5	449	0.56	0.48	5	5	508	0.50	0.46

continued on next page

Table A.1 – continued from previous page

Test	Face grain ∥					Face grain ⊥					End grain				
	σ_N	v	ρ	μ_{stat}	μ_{kin}	σ_N	v	ρ	μ_{stat}	μ_{kin}	σ_N	v	ρ	μ_{stat}	μ_{kin}
2	2.5	5	411	0.51	0.43	2.5	5	498	0.48	0.44	5	5	509	0.51	0.38
3	2.5	5	504	0.50	0.44	2.5	5	504	0.44	0.38	5	5	491	0.45	0.38
4	2.5	5	437	0.47	0.42	2.5	5	431	0.46	0.43	5	5	459	0.52	0.46
5	2.5	5	404	0.43	0.37	2.5	5	477	0.45	0.42	5	5	434	0.46	0.42
MEAN			439	0.51	0.42			472	0.48	0.43			480	0.49	0.42
COV			9%	13%	7%			7%	10%	8%			7%	6%	9%
Punched – perforation 3 and softwood															
1	0.99	5	427	0.78	0.64	0.99	5	-	0.85	0.80	0.99	5	439	0.85	0.66
2	0.99	5	438	0.85	0.69	0.99	5	504	0.67	0.65	0.99	5	499	0.85	0.68
3	0.99	5	502	0.84	0.61	0.99	5	450	0.70	0.61	0.99	5	437	0.84	0.79
4	0.99	5	404	0.69	0.45	0.99	5	444	0.64	0.54	0.99	5	445	0.88	0.71
5	0.99	5	507	0.99	0.74	0.99	5	497	0.69	0.67	0.99	5	435	0.85	0.73
6	0.99	5	437	0.79	0.60	0.99	5	449	0.63	0.62	0.99	5	501	0.93	0.71
7	0.99	5	512	0.75	0.57	0.99	5	412	0.55	0.44	1.00	5	491	0.97	0.79
8	0.99	5	495	0.78	0.62	0.99	5	235	0.67	0.61	0.99	5	447	0.80	0.64
9	0.99	5	411	0.75	0.58	0.99	5	505	0.68	0.64	0.99	5	505	0.88	0.84
10	0.99	5	504	0.75	0.58	0.99	5	502	0.63	0.59	0.99	5	508	0.99	0.85
MEAN			464	0.80	0.61			444	0.67	0.62			471	0.88	0.74
COV			9%	10%	13%			19%	11%	15%			7%	7%	10%
Milled – pyramid pattern 0.5 mm and softwood															
1	2.16	5	428	0.87	0.60	1.70	5	477	1.07	0.46	3.94	5	517	0.72	0.69
2	2.40	5	428	0.82	0.51	1.55	5	475	1.08	0.55	3.85	5	502	0.74	0.68
3	2.04	5	429	0.87	0.69	2.13	5	467	0.96	0.48	3.99	5	491	0.75	0.68
4	2.17	5	432	0.79	0.52	1.87	5	458	0.84	0.61	3.53	5	486	0.82	0.65
5	2.19	5	436	0.88	0.52	1.10	5	454	1.07	0.37	3.03	5	486	0.79	0.70
6	2.14	5	442	0.76	0.49	1.41	5	449	0.82	0.48	3.28	5	478	0.81	0.72
7	2.37	5	442	0.87	0.59	2.32	5	446	0.91	0.36	3.40	5	475	0.78	0.67
8	2.08	5	439	0.80	0.44	2.29	5	445	0.81	0.66	3.54	5	473	0.85	0.70
9	2.22	5	436	0.76	0.51	1.30	5	446	1.03	0.50	3.66	5	471	0.83	0.72
10	2.12	5	438	0.88	0.62	2.44	5	443	0.77	0.52	3.52	5	470	0.84	0.67
11						2.34	5	504	0.73	0.54	1.26	5	383	0.92	0.47

continued on next page

Table A.1 – continued from previous page

Test	Face grain ∥					Face grain ⊥					End grain				
	σ_N	v	ρ	μ_{stat}	μ_{kin}	σ_N	v	ρ	μ_{stat}	μ_{kin}	σ_N	v	ρ	μ_{stat}	μ_{kin}
12						1.89	5	494	0.81	0.53	1.52	5	384	0.91	0.57
13						2.21	5	493	0.63	0.48	1.43	5	389	0.96	0.68
14						2.02	5	487	0.84	0.60	1.45	5	401	1.03	0.59
15						2.14	5	483	0.69	0.63	1.26	5	457	0.93	0.59
16						1.97	5	382	0.82	0.60	1.52	5	467	0.80	0.45
17						1.82	5	381	0.72	0.69	1.34	5	469	0.90	0.52
18						1.56	5	399	0.87	0.57	1.76	5	489	0.77	0.47
19						1.97	5	393	0.85	0.42	1.56	5	574	0.89	0.52
20						1.42	5	385	0.97	0.54	1.70	5	597	0.89	0.53
21						2.17	5	416	0.64	0.67	1.81	5	426	0.72	0.63
22						2.34	5	413	0.75	0.64	1.40	5	430	0.83	0.64
23						2.66	5	410	0.71	0.65	1.87	5	433	0.76	0.68
24						2.35	5	408	0.75	0.60	1.60	5	434	0.78	0.70
25						1.83	5	407	0.94	0.50	1.54	5	435	0.79	0.75
26						1.96	5	552	0.94	0.55	1.58	5	437	0.89	0.77
27						1.22	5	547	1.05	0.49	1.75	5	439	0.75	0.70
28						1.82	5	563	0.87	0.51	1.70	5	240	0.83	0.70
29						2.18	5	559	0.76	0.44	1.71	5	443	0.79	0.75
30						1.81	5	556	1.03	0.52	1.84	5	448	0.71	0.61
31						0.73	5	489	0.94	0.51					
32						0.77	5	452	1.00	0.56					
33						1.37	5	448	0.74	0.61					
34						0.97	5	438	0.95	0.76					
35						1.02	5	433	0.98	0.62					
36						1.01	5	431	0.87	0.67					
37						1.09	5	417	1.00	0.59					
38						0.95	5	415	0.91	0.55					
39						1.28	5	409	0.76	0.75					
40						1.81	5	406	0.83	0.57					
MEAN			435	0.83	0.55			453	0.87	0.56			454	0.83	0.64

continued on next page

Table A.1 – continued from previous page

Test	Face grain ∥					Face grain ⊥					End grain				
	σ_N	v	ρ	μ_{stat}	μ_{kin}	σ_N	v	ρ	μ_{stat}	μ_{kin}	σ_N	v	ρ	μ_{stat}	μ_{kin}
COV			1%	6%	13%			11%	14%	16%			14%	9%	14%
Milled – pyramid pattern 1.0 mm and softwood															
1	2.16	5	376	1.02	0.47	1.53	5	-	0.95	0.78	3.55	5	312	0.64	0.58
2	1.95	5	368	0.92	0.47	1.51	5	-	1.10	0.67	3.39	5	318	0.61	0.64
3	2.06	5	361	0.92	0.45	1.93	5	-	0.87	0.70	1.75	5	320	1.05	0.66
4	2.23	5	353	0.93	0.48	1.81	5	-	0.83	0.59	2.71	5	321	0.76	0.64
5	2.16	5	343	0.95	0.45	2.08	5	-	0.73	0.70	2.70	5	323	0.87	0.69
6	2.03	5	334	1.24	0.50	1.78	5	-	0.92	0.62	2.24	5	326	0.79	0.75
7	2.13	5	328	0.99	0.45	1.52	5	-	1.13	0.55	2.96	5	328	0.73	0.63
8	2.03	5	338	0.98	0.56	1.66	5	-	1.14	0.73	2.44	5	330	0.82	0.76
9	2.08	5	339	0.90	0.44	1.63	5	-	0.89	0.73	2.47	5	336	0.94	0.67
10	1.93	5	322	0.91	0.49	1.84	5	-	1.13	0.68	3.24	5	364	0.83	0.62
11	2.03	5	453	0.73	0.57	1.96	5	-	0.85	0.60					
12	2.14	5	463	0.86	0.63	1.82	5	-	0.93	0.62					
13	2.01	5	467	0.85	0.63	1.92	5	422	1.02	0.74					
14	1.94	5	483	0.87	0.69	2.46	5	426	0.82	0.66					
15	2.27	5	385	0.82	0.57	2.05	5	419	0.86	0.59					
16	2.15	5	400	0.66	0.45										
17	2.44	5	402	0.84	0.56										
18	2.12	5	445	0.92	0.68										
19	2.29	5	468	0.77	0.56										
20	2.19	5	448	0.80	0.66										
MEAN			393	0.89	0.54			422	0.95	0.66			327	0.80	0.66
COV			14%	13%	0.16			1%	14%	0.10			4%	16%	0.08
Milled – pyramid pattern 1.5 mm and softwood															
1	1.96	5	338	0.94	0.43	1.18	5	-	1.19	0.66	1.32	5	-	1.24	0.48
2	1.96	5	339	1.22	0.44	1.50	5	-	1.08	0.52	1.32	5	-	1.02	0.41
3	2.00	5	322	1.16	0.45	1.48	5	-	0.97	0.50	1.37	5	-	1.23	0.39
4	1.93	5	328	1.29	0.39	1.32	5	-	0.99	0.53	1.28	5	-	1.12	0.34
5	1.77	5	334	1.20	0.45	1.25	5	-	1.09	0.62	1.38	5	-	1.14	0.42
6	1.96	5	343	1.18	0.47	1.57	5	-	1.08	0.62	1.31	5	-	1.15	0.48

continued on next page

Table A.1 – continued from previous page

Test	Face grain ∥					Face grain ⊥					End grain				
	σ_N	v	ρ	μ_{stat}	μ_{kin}	σ_N	v	ρ	μ_{stat}	μ_{kin}	σ_N	v	ρ	μ_{stat}	μ_{kin}
7	2.09	5	353	1.21	0.45	1.39	5	-	1.36	0.56	1.43	5	-	1.16	0.44
8	1.97	5	361	0.88	0.33	1.24	5	-	1.12	0.63	1.44	5	-	1.00	0.38
9	1.93	5	368	1.14	0.38	1.38	5	-	0.99	0.46	1.50	5	-	0.92	0.30
10	2.05	5	376	1.08	0.42	1.57	5	-	0.92	0.45	1.46	5	-	1.09	0.33
11	0.64	5	453	0.90	0.62	1.21	5	-	0.99	0.64	1.42	5	-	1.03	0.36
12	1.95	5	463	0.97	0.65	1.63	5	-	1.09	0.62	1.24	5	-	1.09	0.44
13	2.06	5	467	0.88	0.58	1.30	5	-	1.11	0.60	3.10	5	-	0.98	0.61
14	1.80	5	483	0.84	0.51	1.44	5	-	1.08	0.55	3.46	5	-	0.86	0.71
15	1.93	5	385	0.78	0.36	0.84	5	-	0.97	0.51	3.40	5	-	1.01	0.44
16	2.14	5	400	1.07	0.43	0.83	5	-	1.03	0.53	3.10	5	-	0.96	0.44
17	1.92	5	402	1.19	0.51	0.86	5	-	1.15	0.59	2.88	5	-	1.14	0.53
18	2.08	5	445	0.86	0.72	1.06	5	-	0.86	0.55	3.34	5	-	1.00	0.49
19	1.95	5	468	0.94	0.55	0.83	5	-	1.02	0.53					
20	2.35	5	448	0.89	0.55	0.84	5	-	1.22	0.56					
MEAN			393	1.03	0.49			-	1.07	0.56			-	1.06	0.44
COV			0.14	15%	0.21			-	11%	0.11			-	10%	0.23
Milled – pyramid pattern 2.0 mm and softwood															
1						1.32	5	-	1.36	0.81					
2						1.60	5	-	1.13	0.76					
3						1.54	5	-	1.15	0.77					
4						1.55	5	-	1.28	1.09					
5						1.46	5	-	1.30	0.77					
6						1.77	5	-	0.84	0.71					
7						1.53	5	-	1.23	0.87					
8						1.60	5	-	1.14	0.84					
9						1.71	5	-	0.60	0.81					
10						1.47	5	-	1.22	0.83					
11						1.49	5	-	1.08	0.94					
12						1.56	5	-	1.10	0.70					
MEAN								-	1.12	0.83					

continued on next page

Table A.1 – continued from previous page

Test	Face grain ∥					Face grain ⊥					End grain				
	σ_N	v	ρ	μ_{stat}	μ_{kin}	σ_N	v	ρ	μ_{stat}	μ_{kin}	σ_N	v	ρ	μ_{stat}	μ_{kin}
COV								-	19%	0.13					
Milled – circular pattern in DVW and softwood															
1	2.15	5	400	0.87	0.50	1.78	5.00	-	0.80	0.39	2.06	5.00	-	0.78	0.54
2	2.05	5	508	0.69	0.50	2.51	5.00	-	0.98	0.53	1.92	5.00	-	0.63	0.50
3	1.85	5	394	0.86	0.53	2.06	5.00	-	1.06	0.51	1.84	5.00	-	0.93	0.58
4	2.04	5	511	0.69	0.44	1.70	5.00	-	0.85	0.54	1.67	5.00	428	0.84	0.62
5	1.81	5	394	0.92	0.52	2.10	5.00	466	0.94	0.62	1.81	5.00	430	0.94	0.63
6	2.17	5	457	0.84	0.57	1.97	5.00	451	1.03	0.71	1.87	5.00	433	0.76	0.59
7	2.15	5	437	0.73	0.50	2.21	5.00	441	0.79	0.60	1.87	5.00	435	0.81	0.63
8	2.21	5	438	0.78	0.43	1.53	5.00	434	0.97	0.77	1.98	5.00	436	0.80	0.65
9	1.99	5	395	0.89	0.59	1.77	5.00	432	0.96	0.76	1.76	5.00	438	0.88	0.64
10	2.03	5	390	0.68	0.37	2.09	5.00	421	0.77	0.75	1.77	5.00	439	0.86	0.62
11	2.04	5	400	0.83	0.53	1.99	5.00	416	0.98	0.67	1.86	5.00	441	0.77	0.58
12	2.16	5	489	0.70	0.37	1.78	5.00	412	0.88	0.86	1.93	5.00	445	0.81	0.70
13	1.97	5	432	0.81	0.55	2.65	5.00	407	0.61	0.69	1.88	5.00	453	0.86	0.72
14	2.11	5	451	0.74	0.43	1.59	5.00	404	0.83	0.58					
15	2.17	5	417	0.76	0.50										
16	2.05	5	502	0.70	0.47										
17	2.33	5	515	0.80	0.49										
18	2.21	5	438	0.82	0.51										
19	2.13	5	423	0.84	0.54										
20	2.15	5	414	0.69	0.44										
MEAN			440	0.78	0.49			428	0.89	0.64			438	0.82	0.62
COV			10%	10%	0.12			5%	14%	0.20			2%	10%	9%
Milled – circular pattern in steel and softwood															
1	2.42	5	475	0.59	0.46	2.42	5	470	0.60	0.40	4.83	5	492	0.52	0.57
2	2.50	5	428	0.70	0.44	2.50	5	455	0.59	0.51	5.00	5	492	0.55	0.64
3	2.46	5	432	0.59	0.46	2.47	5	451	0.63	0.59	4.93	5	476	0.45	0.42
4	2.50	5	479	0.61	0.42	2.50	5	462	0.71	0.55	5.00	5	476	0.46	0.50
5	2.45	5	517	0.50	0.39	2.45	5	440	0.56	0.45	4.90	5	436	0.60	0.61
6	2.46	5	464	0.56	0.36	2.46	5	403	0.61	0.36	4.93	5	436	0.67	0.51

continued on next page

Table A.1 – continued from previous page

Test	Face grain ‖					Face grain ⊥					End grain				
	σ_N	v	ρ	μ_{stat}	μ_{kin}	σ_N	v	ρ	μ_{stat}	μ_{kin}	σ_N	v	ρ	μ_{stat}	μ_{kin}
7	1.79	5	466	0.54	0.43	1.79	5	460	0.65	0.44	3.58	5	492	0.77	0.54
8	1.86	5	436	0.63	0.42	1.86	5	485	0.66	0.50	3.72	5	492	0.72	0.47
9	1.83	5	472	0.59	0.42	1.82	5	411	0.54	0.37	3.65	5	476	0.62	0.40
10	1.84	5	443	0.57	0.38	1.84	5	470	0.68	0.48	3.67	5	402	0.61	0.47
11	1.86	5	484	0.51	0.39	1.86	5	472	0.67	0.45	3.72	5	436	0.63	0.55
12	1.84	5	483	0.58	0.43	1.83	5	478	0.69	0.47	3.67	5	436	0.77	0.54
13											2.50	5	384	0.80	0.57
14											2.50	5	412	0.78	0.36
15											2.50	5	404	0.47	0.37
16											2.50	5	396	0.67	0.35
17											2.50	5	400	0.81	0.41
18											2.50	5	548	0.68	0.65
19											2.50	5	384	0.71	0.57
20											2.50	5	392	0.54	0.62
21											2.50	5	420	0.61	0.59
22											2.50	5	396	0.60	0.60
23											2.50	5	432	0.64	0.51
24											2.50	5	566	0.58	0.60
MEAN			465	0.58	0.42			455	0.63	0.46			445	0.64	0.52
COV			6%	9%	7%			6%	8%	15%			11%	17%	18%
Milled – horizontal grooves in steel and softwood															
1	2.49	10	398	1.31	0.81	2.49	10	465	0.68	0.56	2.48	10	384	1.09	1.07
2	2.49	10	398	1.04	0.67	2.49	10	460	0.64	0.46	2.48	10	376	1.27	0.81
3	2.49	10	414	1.20	0.65	2.49	10	464	0.66	0.59	2.41	10	586	1.50	1.27
4	2.49	10	474	1.04	0.68	2.48	10	418	0.70	0.58	2.48	10	462	1.21	0.84
5	2.49	10	467	1.03	0.66	2.49	10	459	0.80	0.56	2.48	10	450	1.38	0.82
MEAN			430	1.12	0.69			453	0.70	0.55			452	1.29	0.96
COV			9%	11%	10%			4%	9%	10%			19%	12%	21%
Milled – scale pattern in DVW and softwood															
1	2.49	10	424	0.35	0.25	1.81	5	-	0.59	0.42	2.48	10	429	0.56	0.42
2	2.50	10	424	0.30	0.24	1.82	5	-	0.65	0.38	2.48	10	432	0.64	0.46

continued on next page

Table A.1 – continued from previous page

Test	Face grain ∥					Face grain ⊥					End grain				
	σ_N	v	ρ	μ_{stat}	μ_{kin}	σ_N	v	ρ	μ_{stat}	μ_{kin}	σ_N	v	ρ	μ_{stat}	μ_{kin}
3	2.49	10	426	0.39	0.28	1.83	5	-	0.58	0.36	2.48	10	380	0.59	0.45
4	2.49	10	426	0.30	0.27	1.71	5	-	0.68	0.38	2.48	10	434	0.54	0.42
5	2.49	10	444	0.39	0.29	1.85	5	-	0.67	0.56	2.48	10	394	0.59	0.43
6	2.49	10	444	0.42	0.31	2.17	5	-	0.75	0.65					
7	2.49	10	507	0.35	0.27	2.79	5	-	0.69	0.82					
8	2.49	10	393	0.43	0.33	2.49	10	393	0.50	0.39					
9	2.09	5	-	0.57	0.35	2.49	10	462	0.50	0.39					
10	2.00	5	-	0.53	0.35	2.49	10	425	0.45	0.39					
MEAN			436	0.40	0.29			427	0.61	0.48			414	0.58	0.44
COV			8%	22%	13%			8%	16%	32%			6%	6%	4%
Belt grinding – sanded DVW and softwood															
1						2.47	5	-	0.44	0.26	2.57	5	-	0.31	0.23
2						2.16	5	-	0.54	0.37	2.53	5	-	0.43	0.40
3						2.04	5	-	0.51	0.33	2.63	5	-	0.49	0.39
4						2.05	5	-	0.65	0.48	2.64	5	-	0.55	0.42
5						2.37	5	-	0.58	0.43	2.59	5	-	0.48	0.44
6						1.79	5	-	0.63	0.52	2.65	5	-	0.55	0.45
MEAN								-	0.56	0.40			-	0.47	0.39
COV								-	14%	25%			-	19%	21%
Sandblasting – sandblasted DVW and softwood															
1						2.16	5	-	0.41	0.35	2.29	5	-	0.37	0.35
2						2.26	5	-	0.53	0.42	2.44	5	-	0.48	0.46
3						2.30	5	-	0.41	0.36	2.47	5	-	0.48	0.46
4						2.32	5	-	0.53	0.48	2.50	5	-	0.50	0.47
5						2.13	5	-	0.45	0.44	2.56	5	-	0.55	0.47
6						2.32	5	-	0.59	0.38	2.57	5	-	0.44	0.46
MEAN								-	0.49	0.41			-	0.47	0.44
COV								-	15%	12%			-	13%	10%
Brushing – brushed DVW and softwood															
1						2.5	10	426	0.26	0.21	2.5	10	384	0.44	0.37
2						2.5	10	426	0.27	0.19	2.5	10	384	0.48	0.38

continued on next page

Table A.1 – continued from previous page

Test	Face grain $\parallel$					Face grain $\perp$					End grain				
	σ_N	v	ρ	μ_{stat}	μ_{kin}	σ_N	v	ρ	μ_{stat}	μ_{kin}	σ_N	v	ρ	μ_{stat}	μ_{kin}
3						2.5	10	424	0.26	0.21	2.5	10	417	0.44	0.46
4						2.5	10	424	0.28	0.25	2.5	10	417	0.43	0.40
5						2.5	10	439	0.32	0.29	2.5	10	-	0.56	0.47
MEAN								428	0.28	0.23			401	0.47	0.42
COV								1%	9%	17%			5%	12%	10%
Coated – Sikadur 2K EP and sand and softwood															
1	2.50	5	-	0.83	0.64	2.50	5	-	0.64	0.55	2.50	5	-	0.80	0.74
2	2.50	5	-	0.78	0.72	2.50	5	-	0.70	0.63	2.50	5	-	0.71	0.63
3	2.50	5	-	0.73	0.64	2.50	5	-	0.70	0.65	2.50	5	-	0.54	0.58
4	2.50	5	-	0.67	0.65										
MEAN			-	0.75	0.66			-	0.68	0.61			-	0.69	0.65
COV			-	9%	6%			-	6%	9%			-	19%	13%
Coated – Jowat 2K EP and sand and softwood															
1						1.59	5	-	0.63	0.37	2.23	5	-	0.41	0.34
2						1.78	5	-	0.73	0.49	2.14	5	-	0.53	0.33
3						2.21	5	-	0.56	0.36	1.86	5	-	0.61	0.38
MEAN								-	0.64	0.41			-	0.52	0.35
COV								-	13%	17%			-	19%	7%
Coated – Jowat 2K EP and grit and softwood															
1						1.81	5	-	0.66	0.42	1.39	5	-	0.66	0.50
2						1.87	5	-	0.55	0.36	1.39	5	-	0.73	0.54
3						2.23	5	-	0.60	0.35	1.38	5	-	0.68	0.48
MEAN								-	0.61	0.38			-	0.69	0.51
COV								-	9%	10%			-	5%	6%
Coated – Epoxytape 0.1 mm and sand and softood															
1						2.05	5	-	0.94	0.70	2.39	5	-	0.84	0.66
2						2.20	5	-	0.77	0.61	1.28	5	-	1.02	0.65
3						1.80	5	-	0.75	0.57	1.22	5	-	1.06	0.73
MEAN								-	0.82	0.62			-	0.97	0.68
COV								-	12%	11%			-	12%	6%

continued on next page

Table A.1 – continued from previous page

Test	Face grain ∥					Face grain ⊥					End grain				
	σ_N	v	ρ	μ_{stat}	μ_{kin}	σ_N	v	ρ	μ_{stat}	μ_{kin}	σ_N	v	ρ	μ_{stat}	μ_{kin}
Coated – Epoxytape 1.0 mm and sand and softood															
1						2.28	5	-	0.47	0.39	1.86	5	-	0.79	0.61
2						2.07	5	-	0.75	0.53	1.97	5	-	0.78	0.61
3						2.53	5	-	0.53	0.36	1.60	5	-	0.88	0.61
4						2.66	5	-	0.76	0.64					
5						2.12	5	-	0.71	0.56					
MEAN								-	0.65	0.50			-	0.82	0.61
COV								-	21%	23%			-	6%	1%
Coated – Griptape and softwood															
1						2.13	5	-	0.25	0.13	0.29	5	-	0.39	0.11
2						2.35	5	-	0.22	0.10	2.33	5	-	0.18	0.09
3						2.55	5	-	0.26	0.07	1.17	5	-	0.24	0.09
4											0.53	5	-	0.45	0.09
MEAN								-	0.24	0.10			-	0.32	0.10
COV								-	9%	29%			-	40%	7%

A.2 Connection tests

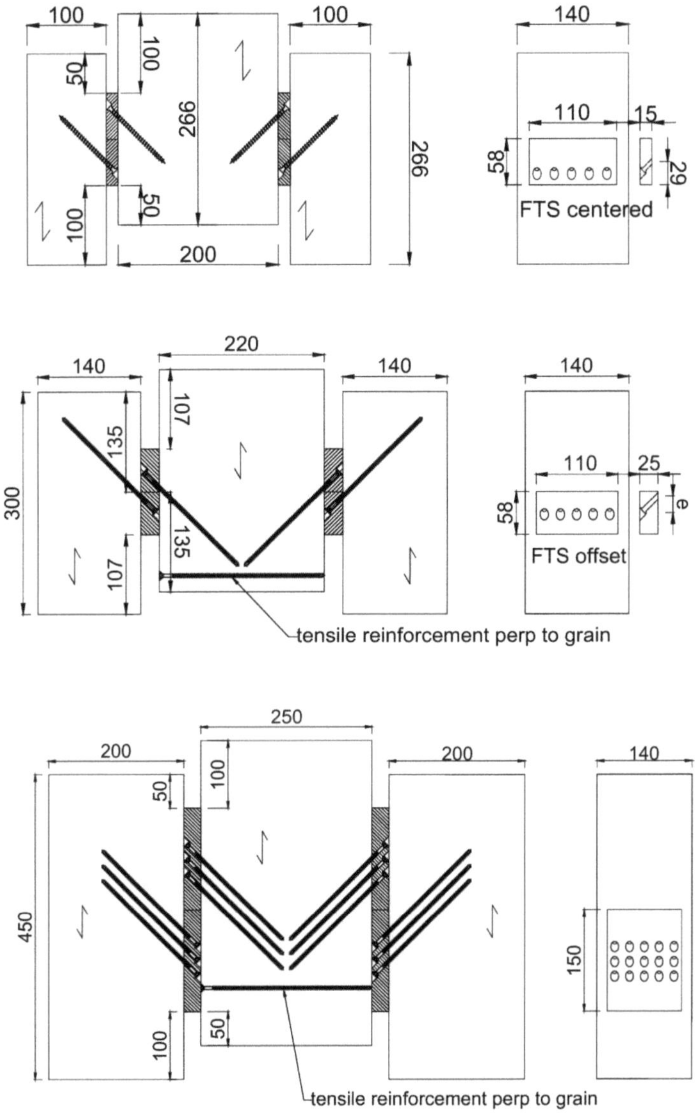

Figure A.2: Test specimens of push-out tests of Series 1–5.

Table A.2: Results of push-out tests of Series 1 (5x 5x100 mm).

Test	$F_{V,test}$ kN	v_{max} mm	K_s N/mm	u %	ρ kg/m^3
Untreated DVW					
1.1	39.7	3.93	17051	11.3	479
1.2	41.2	5.64	16457	11.7	470
1.3	40.7	4.89	16612	11.5	456
MEAN	40.5	4.82	16706	11.5	468
COV	2%	18%	2%	2%	3%
Milled pyramid pattern 1.0 mm					
10.1	55.7	6.89	14946	10.4	468
10.2	48.3	6.43	16467	10.3	479
10.3	54.7	7.63	15439	9.9	483
MEAN	52.9	6.98	15617	10.2	477
COV	8%	9%	5%	3%	2%
Milled pyramid pattern 1.5 mm					
11.1	52.7	6.89	12663	10.0	460
11.2	52.7	8.19	11626	10.5	468
11.3	55.0	7.99	12378	10.4	466
MEAN	53.4	7.69	12222	10.3	465
COV	2%	9%	4%	2%	1%
Milled pyramid pattern 2.0 mm					
12.1	50.5	7.81	11621	10.2	484
12.2	54.6	8.71	11258	10.2	467
12.3	53.3	10.86	9992	10.5	487
MEAN	52.8	9.13	10957	10.3	480
COV	4%	17%	8%	1%	2%
Milled circular pattern					
13.1	50.0	5.44	16284	10.5	484
13.2	50.0	6.34	14245	10.9	488
13.3	49.7	5.98	13628	10.9	487
MEAN	49.9	5.92	14719	10.8	486
COV	0%	8%	9%	2%	0%

continued on next page

Table A.2 – continued from previous page

Test	$F_{V,test}$	v_{max}	K_s	u	ρ
	kN	mm	N/mm	%	kg/m^3
Sanded					
2.1	51.9	7.00	16846	10.5	455
2.2	50.5	7.89	13441	9.6	481
2.3	48.6	5.93	18711	11.0	483
MEAN	50.3	6.94	16333	10.4	473
COV	3%	14%	16%	7%	3%
Sandblasted					
3.1	48.6	5.53	20800	11.6	456
3.2	51.1	5.45	17256	11.3	484
3.3	52.0	6.05	15723	11.1	491
MEAN	50.6	5.67	17926	11.3	477
COV	3%	6%	15%	3%	4%
EpoxyTape (0.1 mm) and sand					
6.1	58.3	7.34	13288	11.4	447
6.2	56.4	7.65	13710	11.6	425
6.3	58.6	7.35	12993	10.9	448
MEAN	57.8	7.45	13330	11.3	440
COV	2%	2%	3%	3%	3%
EpoxyTape (1.0 mm) and sand					
7.1	49.8	6.80	10295	10.6	442
7.2	54.3	8.16	10221	11.3	446
7.3	53.0	7.45	11214	11.3	449
MEAN	52.3	7.47	10577	11.1	446
COV	4%	9%	5%	4%	1%
Griptape					
8.1	36.6	6.19	10879	11.3	423
8.2	40.4	6.73	11655	11.1	451
8.3	40.0	6.73	11792	11.3	440
MEAN	39.0	6.55	11442	11.2	438
COV	5%	5%	4%	1%	3%

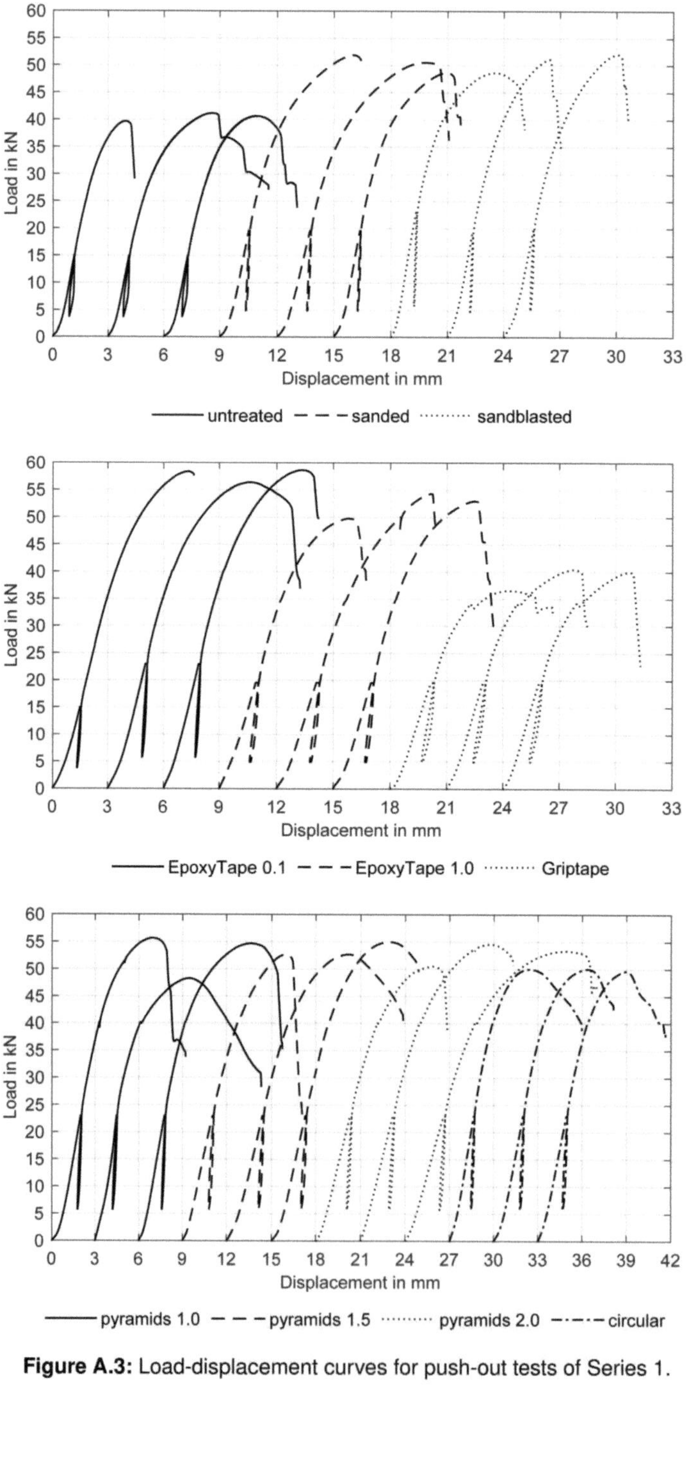

Figure A.3: Load-displacement curves for push-out tests of Series 1.

Table A.3: Results of push-out tests of Series 2 (5x 5x100 mm) and Series 3 (5x 6x180 mm).

Test	$F_{V,test}$ kN	v_{max} mm	K_s N/mm	u %	ρ kg/m³	Failure mode
S2 – milled pyramid pattern 1.0 mm						
1	47.8	5.77	12592	13.0	450	F_{tens}
2	48.9	6.67	11813	12.7	477	F_{tens}
3	47.0	6.72	13706	12.7	457	F_{tens}
4	46.0	4.75	15863	11.2	445	F_{tens}
5	55.7	6.30	16507	12.7	449	F_{ax}
MEAN	49.1	6.04	14096	12.5	455	
COV	8%	13%	14%	6%	3%	
S2 – milled pyramid pattern 1.5 mm						
1 [1)]	10.9	29.4	8061	11.2	464	Error
2	49.0	5.33	17386	11.3	470	F_{tens}
3	41.4	4.74	12899	11.4	453	F_{tens}
4	54.4	5.36	14710	12.3	449	F_{tens}
5	46.4	4.35	15158	12.9	468	F_{tens}
MEAN	47.8	4.95	15038	12.0	460	
COV	11%	10%	12%	6%	2%	
S3 – milled pyramid pattern 0.5 mm						
1	88.8	15.8	19058	12.1	427	F_{tens}
2	87.1	12.2	16992	11.8	444	F_{tens}
3	84.1	11.3	16806	11.9	450	F_{tens}
4	75.5	6.63	19259	12.0	450	F_{tens}
5	86.9	9.86	17052	11.9	455	F_{tens}
MEAN	84.5	11.16	17833	11.9	445	
COV	6%	30%	7%	1%	2%	
S3 – milled circular pattern						
1	79.5	8.24	15010	11.9	466	F_{tens}
2	79.5	9.67	17660	11.8	454	F_{tens}
3	77.9	7.98	19074	11.7	459	F_{tens}
4	83.4	9.75	19330	11.7	465	F_{tens}
5	79.5	7.95	18579	11.9	477	F_{tens}
MEAN	80.0	8.72	17931	11.8	464	
COV	3%	10%	10%	1%	2%	

[1)] erroneous specimen, screws under compressive load

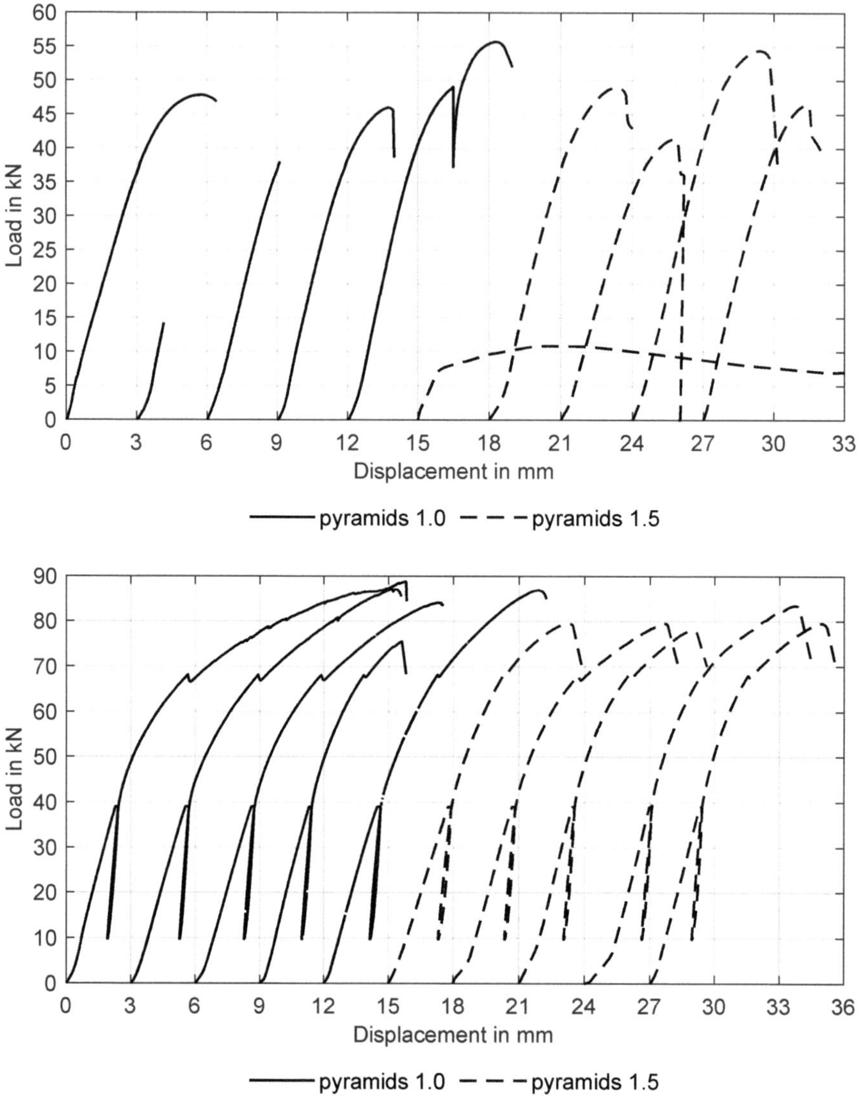

Figure A.4: Load-displacement curves for push-out tests of Series 2–3.

Table A.4: Results of push-out tests of Series 4 (15x 6x100 mm) and Series 5 (15x 6x200 mm).

Test	$F_{V,test}$ kN	v_{max} mm	K_s N/mm	u %	ρ kg/m³	Failure mode
S4 – untreated DVW						
1	122	7.17	37476	12.2	453	F_{ax}
2	122	7.51	33265	12.2	459	F_{ax}
3	116	6.46	31628	12.2	446	F_{ax}
4	119	7.22	32943	12.1	443	F_{ax}
5	117	6.54	32421	12.1	445	F_{ax}
MEAN	119	6.98	33547	12.2	449	
COV	2%	7%	7%	1%	1%	
S4 – milled pyramid pattern 0.5 mm						
1	152	6.60	36751	11.8	453	F_{ax}
2	153	7.60	35346	12.4	445	F_{ax}
3	154	7.44	32233	12.0	442	F_{ax}
4	159	7.73	33279	11.7	460	F_{ax}
5	148	7.06	32533	12.2	465	F_{ax}
MEAN	153	7.29	34028	12.0	453	
COV	3%	6%	6%	2%	2%	
S4 – embossed pattern						
1	143	7.98	43204	12.6	440	F_{ax}
2	138	4.88	52789	12.1	440	F_{ax}
3	144	6.78	40005	12.0	434	F_{ax}
4	139	5.40	43907	12.0	442	F_{ax}
5	138	6.36	33264	12.3	464	F_{ax}
MEAN	140	6.28	42634	12.2	444	
COV	2%	19%	17%	2%	3%	
S5 – embossed pattern						
1	198	11.5	41742	11.4	424	DVW
2	176	13.3	36166	11.4	416	DVW
3	181	10.2	34335	11.5	446	DVW
MEAN	185	12	37414	11.4	429	
COV	6%	13%	10%	0%	4%	

1) error during data acquisition

2) preloaded up to 89.1 kN

3) specimen with reinforcement screws perpendicular to the grain in joist

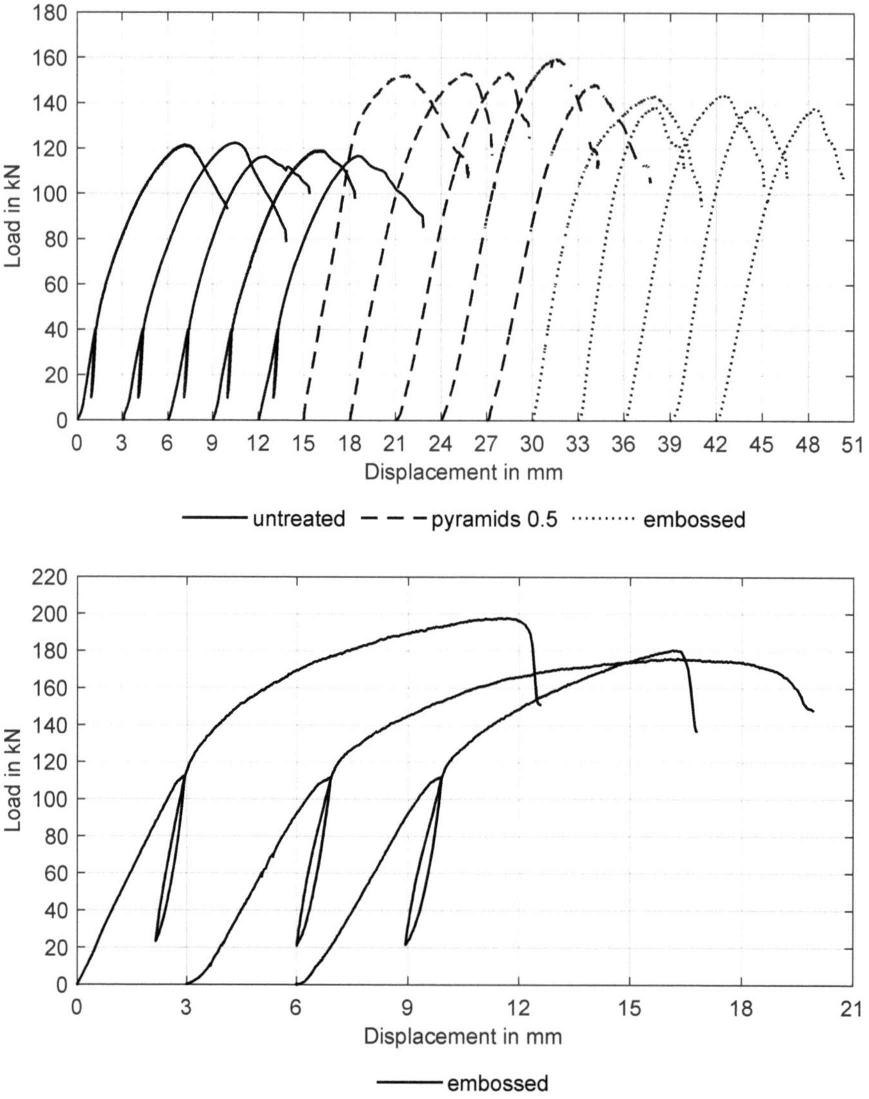

Figure A.5: Load-displacement curves for push-out tests of Series 4–5.

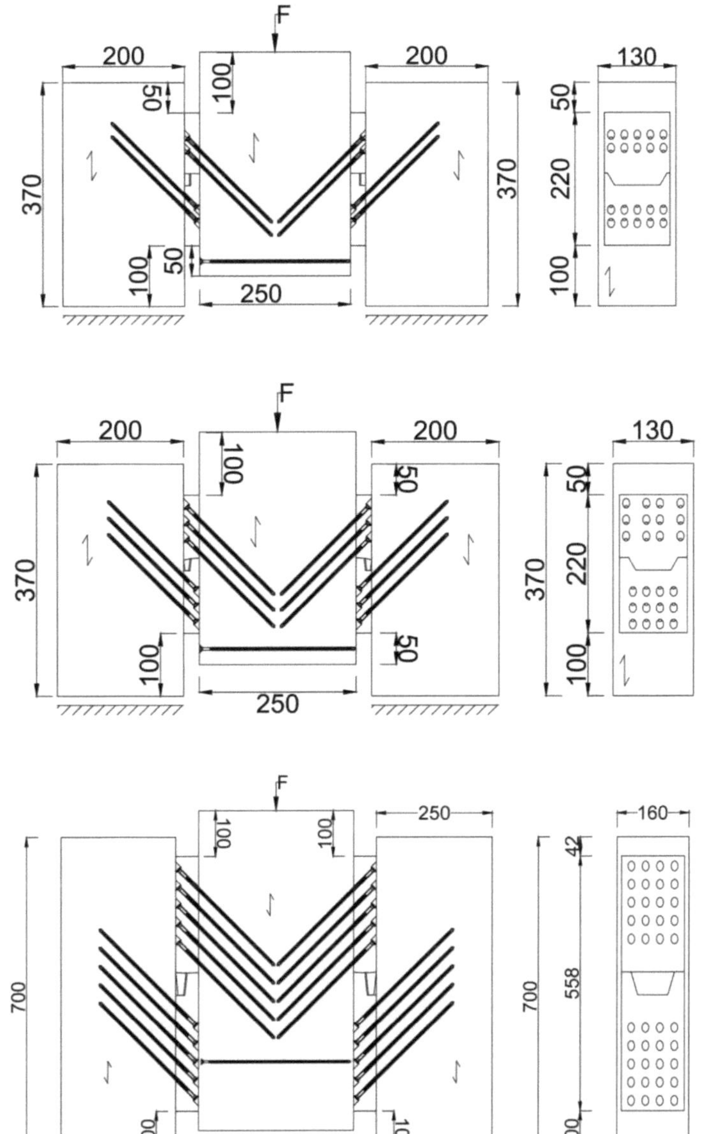

Figure A.6: Test specimens for push-out tests of Series 6–7.

Table A.5: Results of push-out tests of Series 6 (v1 10x 6x000 mm), Series 7 (v2 12x 6x200 mm) and Series 8 (v3 20x 8x300 mm).

Test	$F_{V,test}$ kN	v_{max} mm	K_s N/mm	u %	ρ kg/m³	Failure mode
S6 – milled pyramid pattern 0.5 mm						
1	156	10.8	22043	10.9	471	F_{tens}
2	154	12.8	28319	11.3	458	F_{tens}
3	146	13.8	25598	11.3	447	DVW
4	149	13.3	27749	11.4	445	F_{tens}
5	153	11.7	27762	11.4	464	F_{tens}
10	142	15.0	28220	10.9	450	F_{tens}
11	148	15.3	25611	11.0	435	DVW
12	150	14.9	26282	11.0	441	F_{tens}
MEAN	150	13.4	26448	11.2	451	
COV	3%	12%	8%	2%	3%	
S7 – milled pyramid pattern 1.0 mm						
1 [1)]	119	4.47	49658	11.1	469	Error
2	155	12.0	32666	10.8	445	$F_{c,90}$
3	157	13.7	27952	11.5	440	$F_{c,90}$
4	179	10.7	32053	11.1	489	F_{tens}
5	167	8.27	34577	11.3	463	F_{tens}
6 [2)]	192	9.07	60989	11.8	440	F_{tens}
7 [2)]	180	7.76	65945	11.7	446	$F_{t,90}$ / $F_{v,R}$
MEAN	172	10	42364	11.4	454	
COV	9%	23%	39%	3%	4%	
S8 – milled pyramid pattern 1.0 mm						
1	491	15.6	84983	11.6	431	F_{tens}
2	499	16.8	76199	11.8	466	DVW
3	499	16.8	81385	11.5	424	DVW
4	443	14.1	82032	11.6	458	$F_{c,90}$
5 [2)]	503	9.84	104606	10.6	444	$F_{t,90}$ / $F_{v,R}$
MEAN	487	14.6	85841	11.4	445	
COV	5%	20%	13%	4%	4%	

[1)] error during test execution

[2)] specimen with CLT

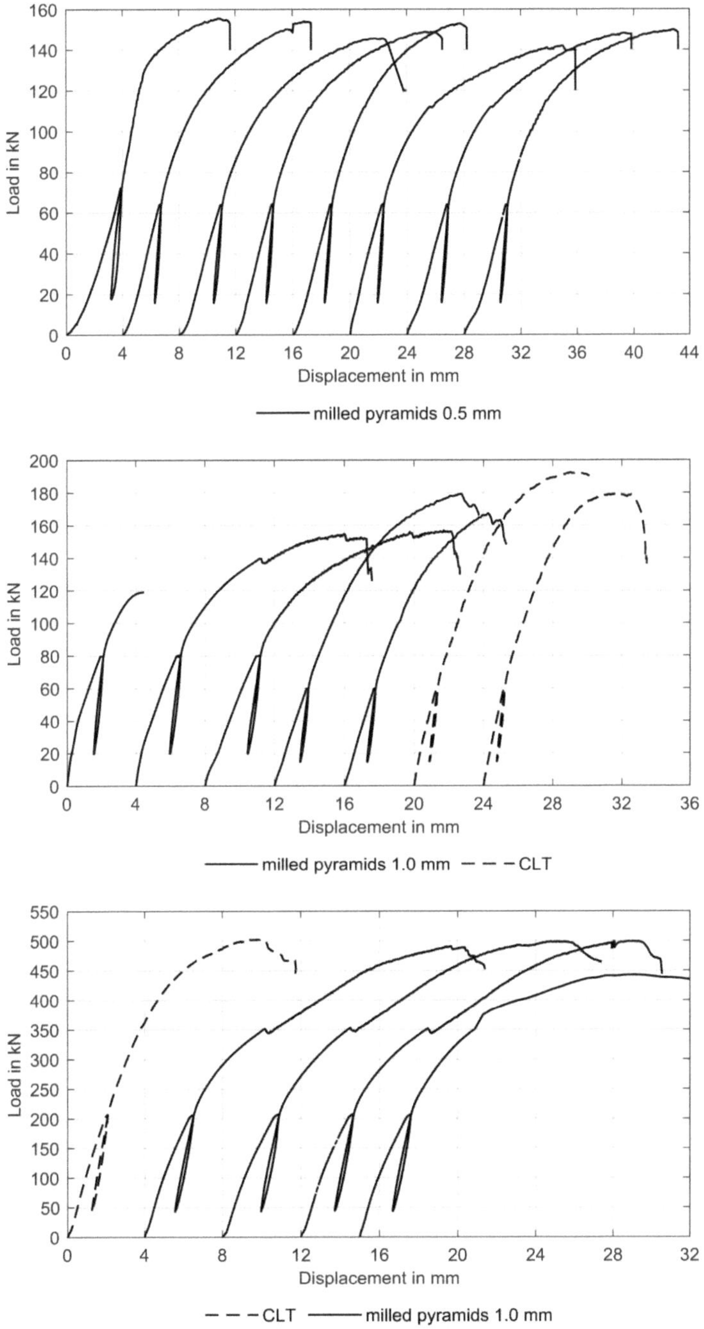

Figure A.7: Load-displacement curves for push-out tests of Series 7–8.

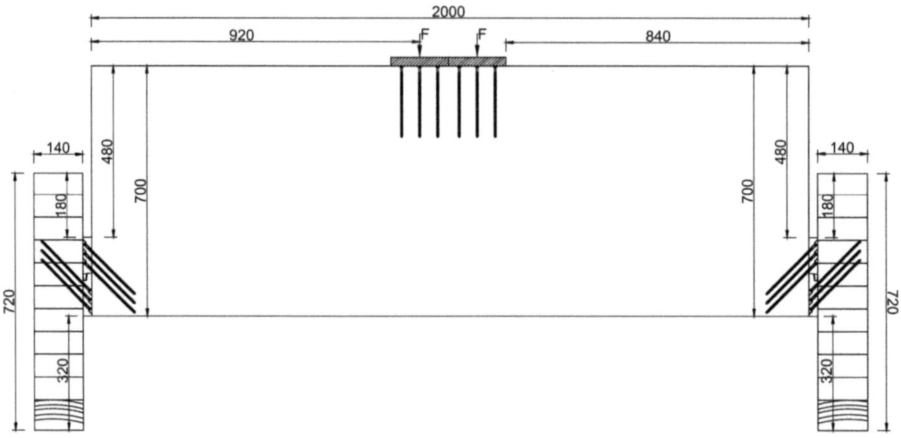

front view

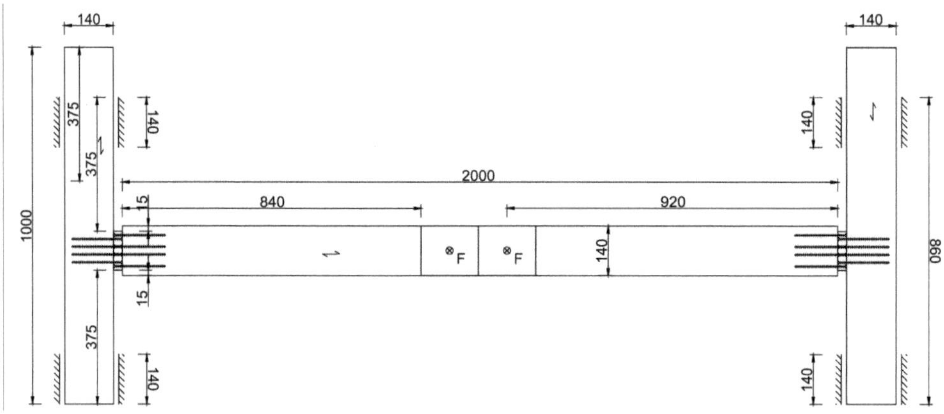

top view

Figure A.8: Test specimens for main beam to secondary beam tests with connector v2.

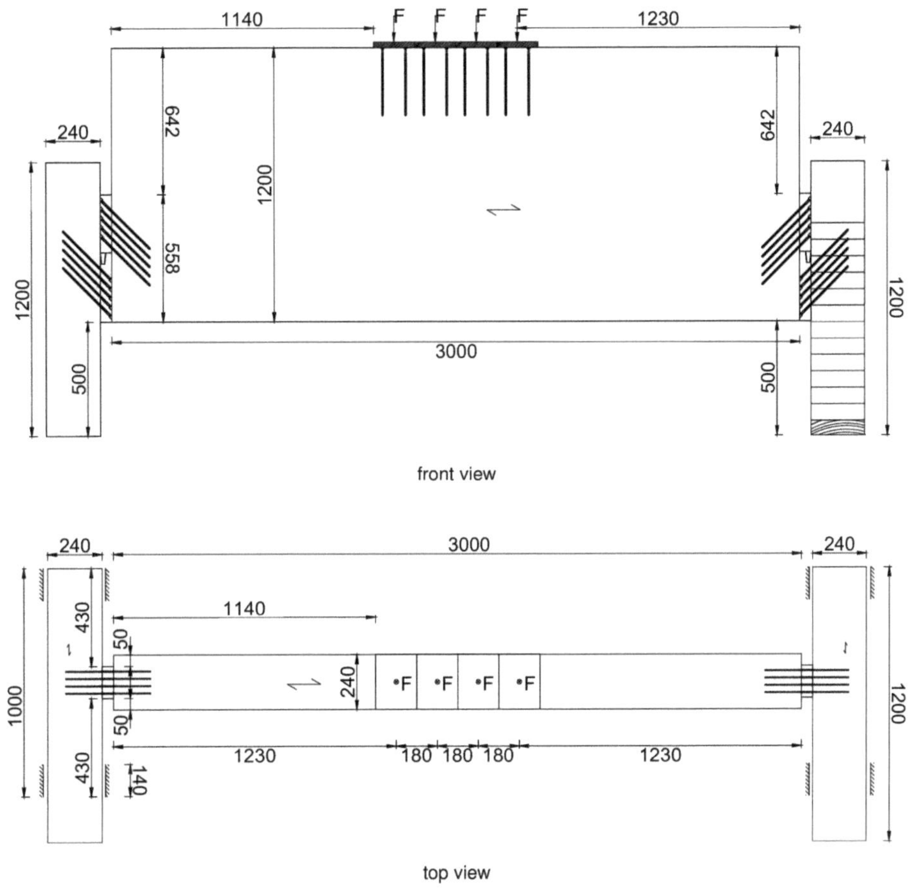

Figure A.9: Test specimens for main beam to secondary beam tests with connector v3.

Table A.6: Results of main beam to secondary beam tests with connector v2 (12x 6x200 mm) and connector v3 (20x 8x300 mm).

Test	$F_{V,test}$ kN	v_{max} mm	K_s N/mm	u %	ρ kg/m³	Failure mode
Connector v2 – milled pyramid pattern 1.0 mm						
0 [1)]	89.1	3.37	27850	12.4	452	Error
1 [2)]	181	10.4	36079	12.4	452	$F_{c,90}$ / F_{ax}
2	165	11.7	28741	12.4	442	$F_{c,90}$ / F_{ax}
3	171	12.9	34065	12.4	459	$F_{c,90}$ / F_{ax}
4	169	11.2	33596	12.5	454	$F_{c,90}$ / F_{ax}
MEAN	172	11.5	33120	12.4	452	
COV	4%	9%	9%	0%	2%	
Connector v3 – milled pyramid pattern 1.0 mm						
1	393	13.0	71021	10.5	429	$F_{t,90}$
2	362	9.89	72778	10.5	435	$F_{t,90}$
3 [3)]	429	12.0	71089	10.6	434	$F_{t,90}$
MEAN	394	11.6	71629	10.5	433	
COV	8%	14%	1%	0%	1%	

1) error during data acquisition

2) preloaded up to 89.1 kN

3) specimen with reinforcement screws perpendicular to the grain in joist

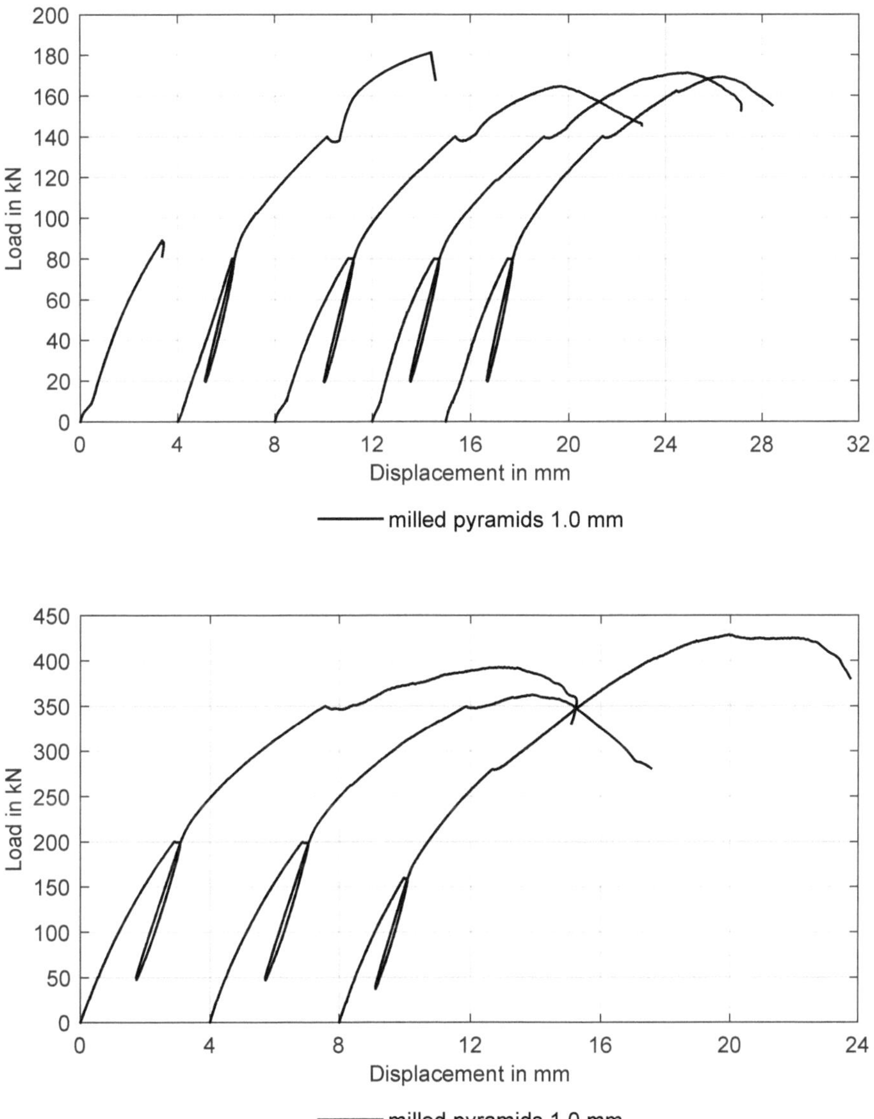

Figure A.10: Load-displacement curves for main beam to secondary beam tests.

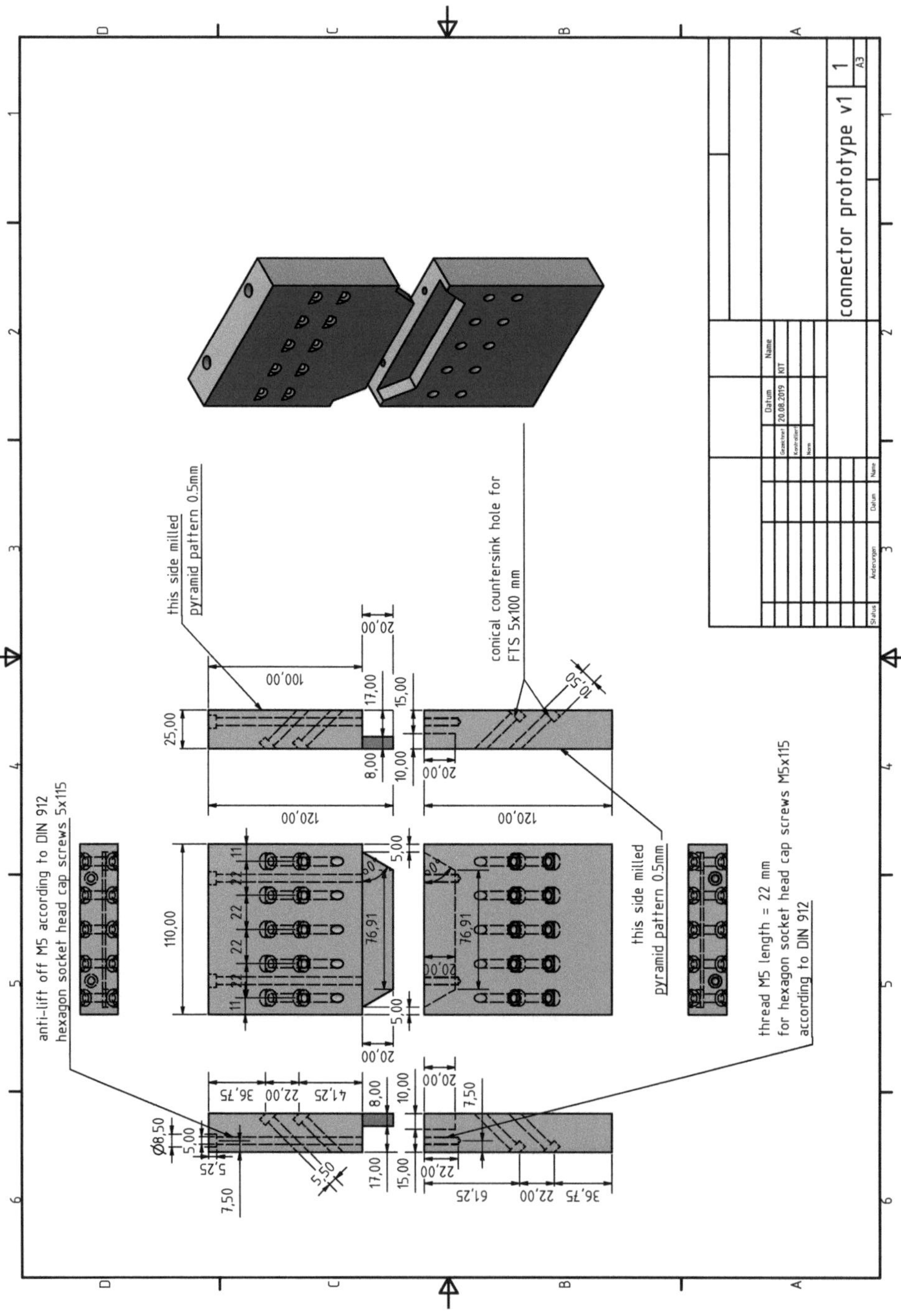

Figure A.11: Detailed drawing of connector v1.

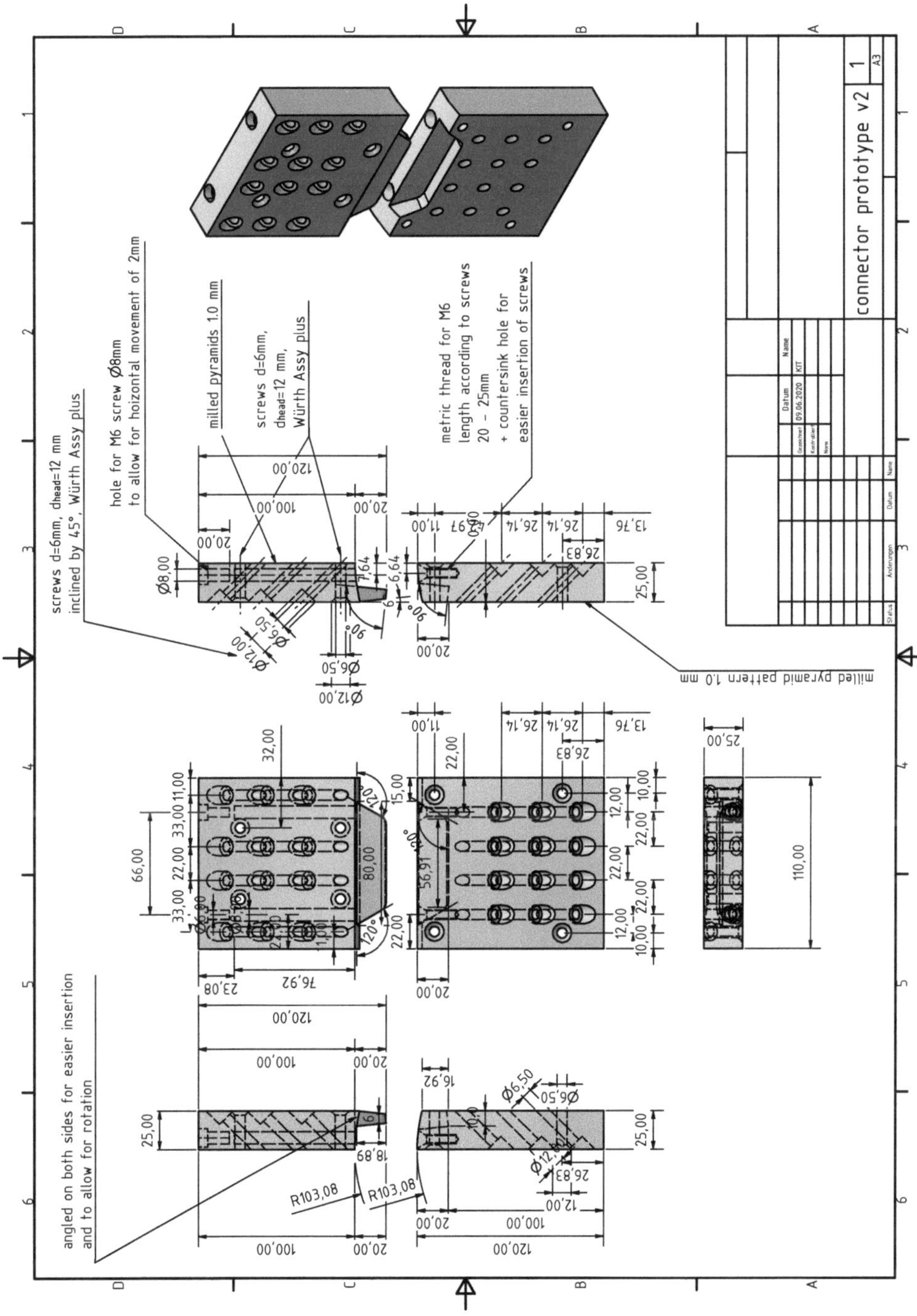

Figure A.12: Detailed drawing of connector v2.

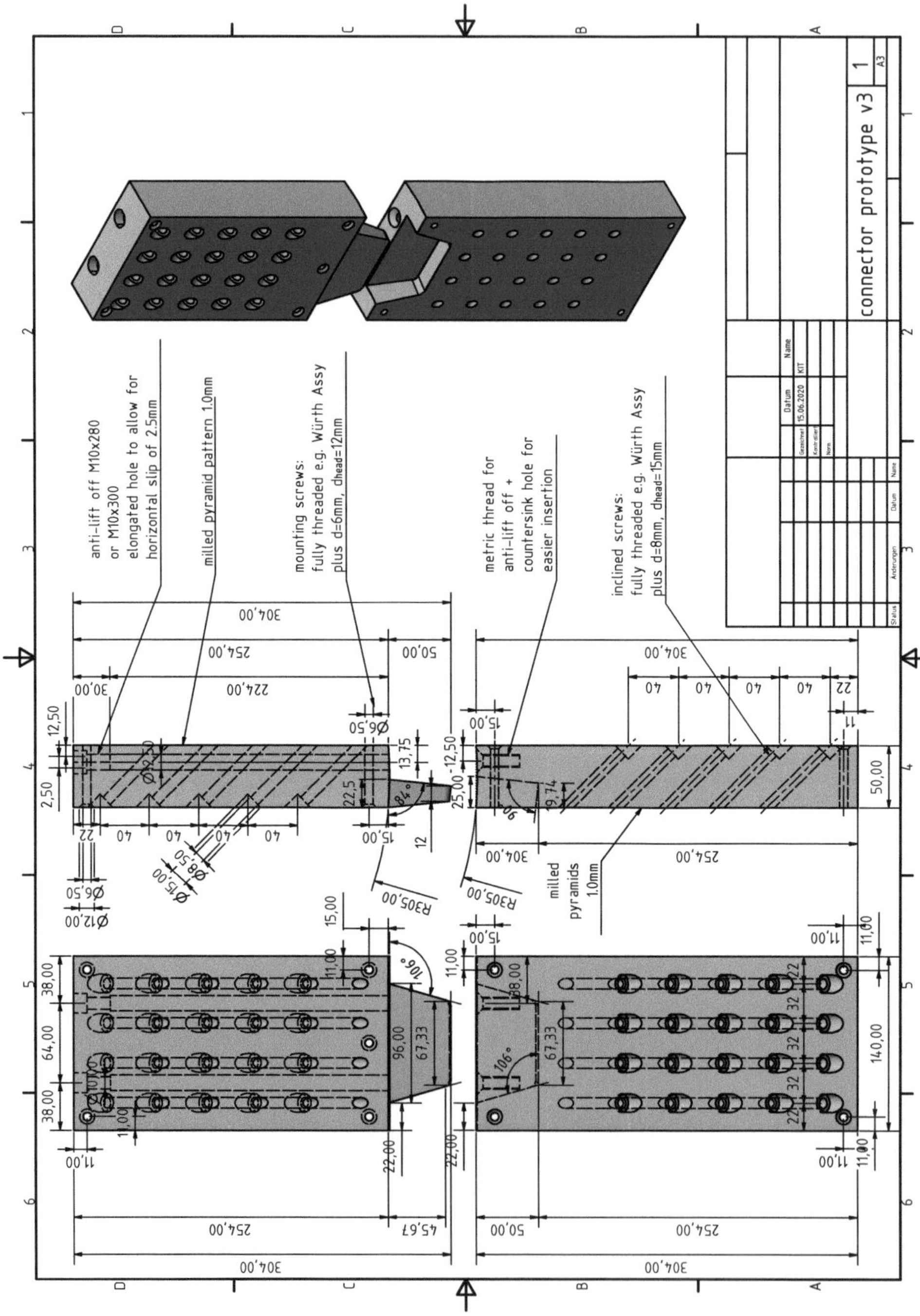

Figure A.13: Detailed of connector v3.

A.3 Long-term tests

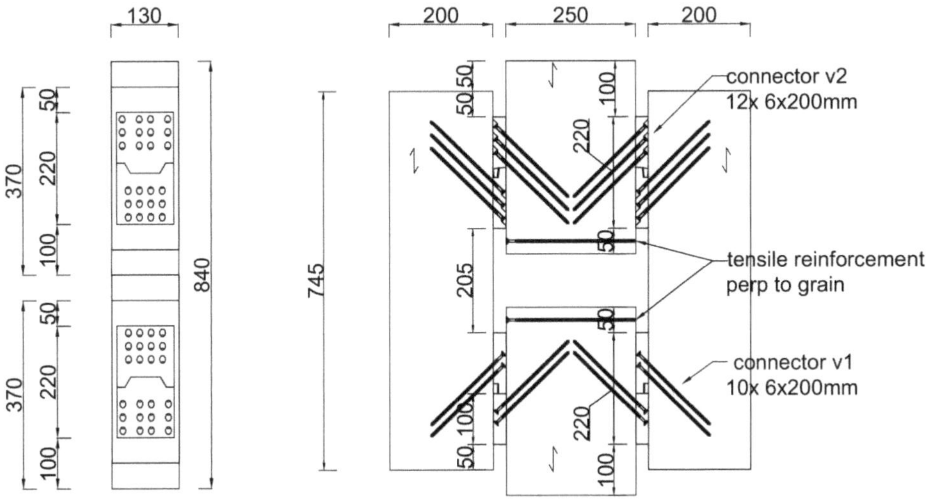

Figure A.14: Test specimens for long-term tests in SC1 and SC2.

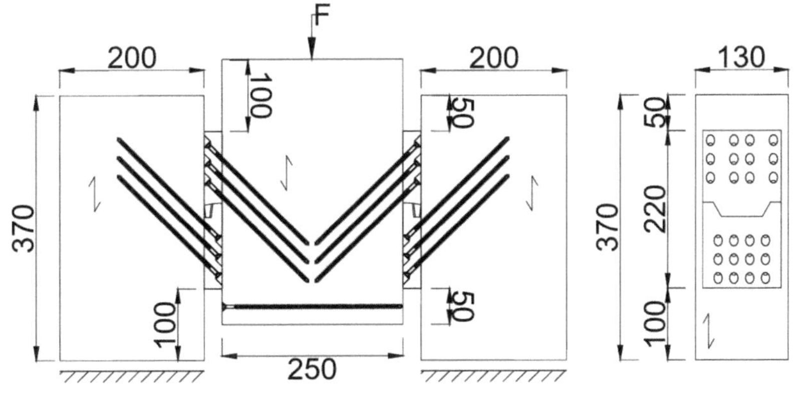

Figure A.15: Test specimens for long-term tests with changing moisture content.

Table A.7: Residual load-carrying capacity of connector v1 (10x 6x200 mm) with milled pyramid pattern.

Test	$F_{V,test}$ kN	v_{max} mm	K_s N/mm	u %	ρ kg/m^3	Failure mode
Service class 1						
1	143	10.7	28248	10.9	436	F_{tens}
Service class 2						
1	139	13.6	28211	15.8	457	DVW

Table A.8: Residual load-carrying capacity of connector v2 (12x 6x200 mm) with milled pyramid pattern.

Test	$F_{V,test}$ kN	v_{max} mm	K_s N/mm	u %	ρ kg/m^3	Failure mode
Service class 1						
1	149	9.91	35014	11.0	447	F_{tens}
2	141	10.8	39016	11.0	438	F_{tens}
3	145	12.5	33089	10.8	434	F_{tens}
MEAN	145	11.1	35706	11.0	440	
COV	3%	12%	8%	1%	1%	
Service class 2						
1	153	12.0	32386	15.3	479	DVW
2	155	12.0	40003	15.0	472	F_{tens}
3	154	11.5	36783	15.3	456	F_{tens}
MEAN	154	11.8	36391	15.2	469	
COV	0%	3%	11%	1%	2%	

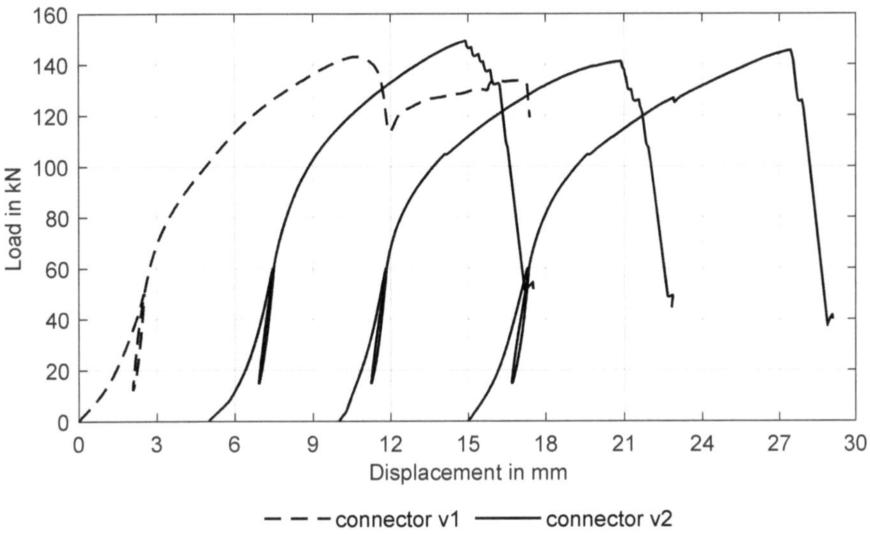

Figure A.16: Load-displacement curves for long-term tests in service class 1.

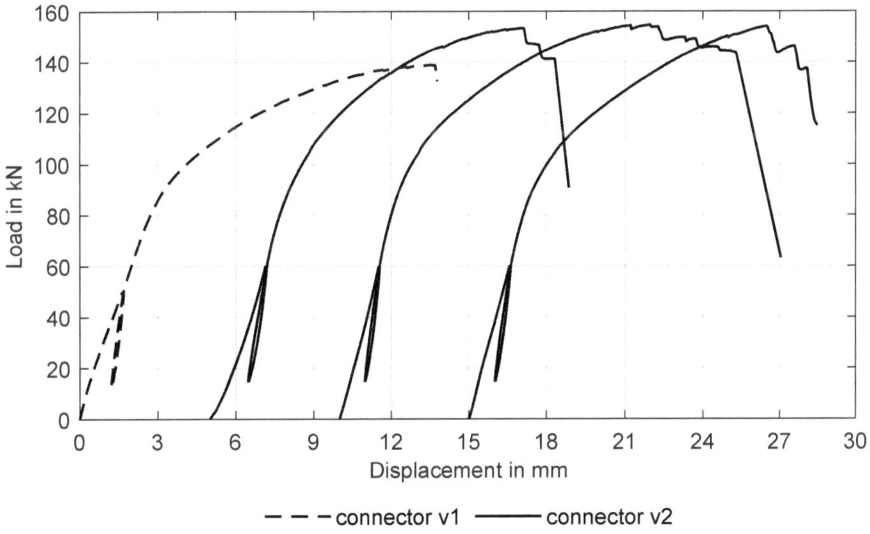

Figure A.17: Load-displacement curves for long-term tests in service class 2.

Table A.9: Residual load-carrying capacity of connector v1 (10x 6x200 mm) with milled pyramid pattern.

Test	$F_{V,test}$ kN	v_{max} mm	K_s N/mm	u %	ρ kg/m^3	Failure mode
m.c. = 12% → 16% → 12%						
1	145	12.9	23893	11.1	470	F_{tens}
2	135	12.7	21932	11.3	446	F_{tens}
3	146	18.3	22713	11.3	431	F_{tens}
MEAN	142	14.7	22846	11.2	449	
COV	4%	22%	4%	1%	4%	

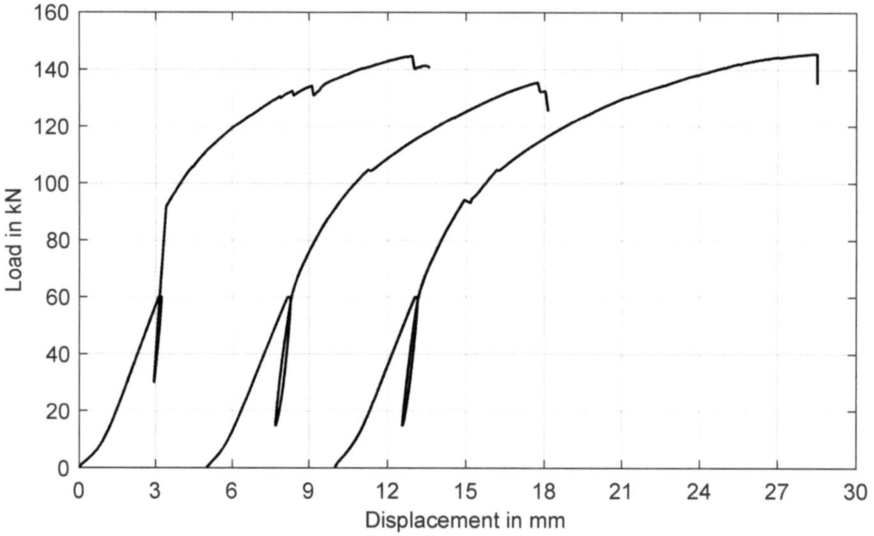

Figure A.18: Load-displacement curves for long-term tests with changing moisture content.

A.4 Modelling

Table A.10: Engineering constants for elastic material properties.

	2D/3D DVW	3D (Series 1) wood	3D (v3) wood	3D soft material	3D steel
E1	13 797	11 000	12 100	370	210 000
E2	15 848	370	300	50	
E3	15 848	370	300	370	
Nu12	0	0.46	0.46	0	0.3
Nu13	0	0.45	0.45	0	
Nu23	0	0.5	0.5	0	
G12	1 400	690	650	690	
G13	1 400	690	650	690	
G23	560	69	65	69	

Table A.11: Spring properties for screws of connector v2 for 2D-model.

*Connector Behavior, name=inclined		*Connector Behavior, name=normal	
*Connector Elasticity, nonlinear, component=1		*Connector Elasticity, nonlinear, component=1	
0	0	0	0
45120	0.234	22560	0.23
53580	0.617	26790	0.615
56400	1	28200	1
*Connector Elasticity, nonlinear, component=2		*Connector Elasticity, nonlinear, component=2	
-5640.1	-10	-2820.1	-10
-5640	-1	-2820	-1
-5358	-0.617	-2679	-0.615
-4512	-0.234	-2256	-0.23
4512	0.234	2256	0.23
5358	0.617	2679	0.615
5640	1	2820	1
5640.1	10	2820.1	10
*Connector Failure, component=1		*Connector Failure, component=1	
Lower bound	Upper bound	Lower bound	Upper bound
-1	56401	-1	28201

Table A.12: Spring properties for compression perpendicular to the grain of wood for 2D-model.

Connector v2		Connector v3	
*Connector Behavior, name=perpendicular			
*Connector Elasticity, nonlinear, component=1			
-7938	-47	-15033.6	-47
-7937.97	-30	-15033.5	-30
-7937.76	-25	-15033.1	-25
-7936.21	-20	-15029.9	-20
-7935.32	-19	-15028.2	-19
-7934.01	-18	-15025.5	-18
-7932.05	-17	-15021.5	-17
-7929.12	-16	-15015.5	-16
-7924.75	-15	-15006.6	-15
-7918.23	-14	-14993.3	-14
-7908.51	-13	-14973.5	-13
-7894	-12	-14944	-12
-7872.37	-11	-14899.9	-11
-7840.08	-10	-14834.2	-10
-7791.93	-9	-14736.1	-9
-7720.09	-8	-14589.8	-8
-7612.91	-7	-14371.6	-7
-7453.02	-6	-14046	-6
-7214.5	-5	-13560.3	-5
-6795.11	-4.5	-12819.4	-4.5
-6340.26	-4	-12006.2	-4
-5842.09	-3.5	-11104.9	-3.5
-5291.02	-3	-10095.8	-3
-4675.32	-2.5	-8955.12	-2.5
-3980.69	-2	-7653.71	-2
-3189.65	-1.5	-6155.98	-1.5
-2280.87	-1	-4418.47	-1
-1228.27	-0.5	-2388.09	-0.5
0	0	0	0
0.1	100	0.1	100
*Connector Friction, component=2, contact force=1			
0		0	
*Friction			
0.94		0.94	

Table A.13: Spring properties for compression parallel to the grain of wood for 2D-model.

Connector v2		Connector v3	
*Connector Behavior, name=parallel			
*Connector Elasticity, nonlinear, component=1			
-26400	-2	-47040	-3
0	0	0	0
0.1	100	0.1	100
*Connector Friction, component=2, contact force=1			
0		0	
*Friction			
0.8		0.8	

Table A.14: Spring properties for screws of connector v3 for 2D-model.

*Connector Behavior, name=inclined		*Connector Behavior, name=normal	
*Connector Elasticity, nonlinear, component=1		*Connector Elasticity, nonlinear, component=1	
0	0	0	0
77120	0.316	22560	0.249
91580	0.658	26790	0.625
96400	1	28200	1
*Connector Elasticity, nonlinear, component=2		*Connector Elasticity, nonlinear, component=2	
-9640.1	-10	-2820.1	-10
-9640	-1	-2820	1
-9158	-0.658	-2679	-0.625
-7712	-0.316	-2256	-0.249
0	0	0	0
7712	0.316	2256	0.249
9158	0.658	2679	0.625
9640	1	2820	1
9640.1	10	2820.1	10
*Connector Failure, component=1		*Connector Failure, component=1	
Lower bound	Upper bound	Lower bound	Upper bound
-1	96401	-1	28201

Table A.15: Input parameters for cohesive surface for 3D-model.

	*Cohesive Behavior		
	Knn	Kss	Ktt
Series 1	0	31.91	31.91
Series 1 - fitted	0	31.91	31.91
Connector v3	0	21.224	21.224
	*Damage Initiation, criterion=MAXS (maximum nominal stress)		
	Normal only	Shear-1 Only	Shear-2 Only
Series 1	100	12	12
Series 1 - fitted	100	5	5
Connector v3	100	5	5
	*Damage Evolution, type=ENERGY		
	Fracture Energy		
All	35		
	*Damage Stabilization		
All	0.0001		